T0318543

Ultrasound Guided Musculoskeletal Procedures in Sports Medicine

Ultrasound Guided Musculoskeletal Procedures in Sports Medicine

A Practical Atlas

Dinesh Sirisena

MRCGP(UK) | Dip(SEM)(UK&I) | PgD(MedUSS)(UK) |
MSc(SEM)(UK) | MSc(Clin. Ed.)(UK) | FFSEM(Ire) |
FFSEM(UK&I) | FAMS(SEM)(Sg)

ACADEMIC PRESS
An imprint of Elsevier

ELSEVIER

Academic Press is an imprint of Elsevier
125 London Wall, London EC2Y 5AS, United Kingdom
525 B Street, Suite 1650, San Diego, CA 92101, United States
50 Hampshire Street, 5th Floor, Cambridge, MA 02139, United States
The Boulevard, Langford Lane, Kidlington, Oxford OX5 1GB, United Kingdom

Copyright © 2021 Elsevier Inc. All rights reserved.

No part of this publication may be reproduced or transmitted in any form or by any means, electronic or mechanical, including photocopying, recording, or any information storage and retrieval system, without permission in writing from the publisher. Details on how to seek permission, further information about the Publisher's permissions policies and our arrangements with organizations such as the Copyright Clearance Center and the Copyright Licensing Agency, can be found at our website: www.elsevier.com/permissions.

This book and the individual contributions contained in it are protected under copyright by the Publisher (other than as may be noted herein).

Notices
Knowledge and best practice in this field are constantly changing. As new research and experience broaden our understanding, changes in research methods, professional practices, or medical treatment may become necessary.

Practitioners and researchers must always rely on their own experience and knowledge in evaluating and using any information, methods, compounds, or experiments described herein. In using such information or methods they should be mindful of their own safety and the safety of others, including parties for whom they have a professional responsibility.

To the fullest extent of the law, neither the Publisher nor the authors, contributors, or editors, assume any liability for any injury and/or damage to persons or property as a matter of products liability, negligence or otherwise, or from any use or operation of any methods, products, instructions, or ideas contained in the material herein.

Library of Congress Cataloging-in-Publication Data
A catalog record for this book is available from the Library of Congress

British Library Cataloguing-in-Publication Data
A catalogue record for this book is available from the British Library

ISBN: 978-0-323-91014-9

For information on all Academic Press publications visit our website at
https://www.elsevier.com/books-and-journals

Publisher: Stacy Masucci
Acquisitions Editor: Elizabeth Brown
Editorial Project Manager: Pat Gonzalez
Production Project Manager: Kiruthika Govindaraju
Cover Designer: Alan Studholme

Typeset by TNQ Technologies

Contents

Biography

Dr. Dinesh Sirisena is a Sports and Exercise Medicine (SEM) Consultant based at the Khoo Teck Puat Hospital (KTPH) Sports Medicine Centre. After graduating from Bart's and the London Medical School, he completed his SEM training in London before moving to, Singapore. While he sees and treats a wide range of sports-related and general MSK conditions, his areas of interest include hip, spine and shoulder injuries. In addition to providing rehabilitation and focussed shockwave therapies, Dr Sirisena is also an interventional musculoskeletal sonographer, providing diagnostic ultrasound scanning together with specialised procedures such as  ultrasound guided PRP/Prolotherapy/high volume/spinal/peripheral nerve injections and tendon stripping. To go with this, he has written a book titled "Ultrasound Guided Musculoskeletal Procedures in Sports Medicine: A Practical Atlas."

In Asia, Dr Sirisena is the nominated physician for the European Professional Golf Association, but in addition, he has worked with a number of elite and mass participation sports. He provided medical services for the Football Association of Singapore, Singapore Rugby, the Standard Chartered Marathon and other sporting events. He was the team physician for Team Singapore at the 2017 South East Asian (Kuala Lumpur, Malaysia) and 2018 Asian (Palembang, Indonesia) games.

Dr Sirisena is Adjunct Assistant Professor for the Yong Loo Lin (National University of Singapore) and Lee Kong Chian (Nanyang Technological University) medical schools. He has also been involved in a number of research projects, with several peer reviewed publications in notable journals. Indeed, one of the key motivations for him is to help his peers and students develop their own research interests and inquisitiveness.

Follow him on Twitter (@sports_med_doc) or Instagram (@sport_med_doc) and remember to check out www.sportsmedinfo.sg a free resource for clinicians and patients alike.

Foreword

I am honoured to contribute the Foreword for this excellent book *Ultrasound Guided Musculoskeletal Procedures in Sports Medicine*. Written by Dr Dinesh Sirisena, it serves its purpose as an effective and practical guide for anyone interested in learning to perform ultrasound guided procedures in patients with musculoskeletal sports injuries. This book will also be a valuable *aide-memoire* for experienced practitioners seeking a handy reference source.

I first encountered Dinesh 5 years ago, when he joined the Sports Medicine Centre of Khoo Teck Puat Hospital. This Centre is one of the most established Sports Medicine Centres in Singapore, with luminaries such as Dr Teh Kong Chuan and Dr Teoh Chin Sim. To say that Dinesh burst onto the scene is almost an understatement — though relatively young, he quickly established himself on the local and regional sports medicine scene. Besides his clinical work in sports at all levels, he is an enthusiastic speaker, teacher and author, as well as a dynamic collaborator with a spectrum of medical and healthcare professionals, including the musculoskeletal radiology community. Indeed, with his move here, the UK's loss is Singapore's gain.

Ultrasound Guided Musculoskeletal Procedures in Sports Medicine is a comprehensive book that covers — literally from neck to toe — all possible injections for musculoskeletal pathologies caused by sports injuries. How exactly to perform each procedure is described in detail. Among the great strengths of this book are its beautiful illustrations — consisting of anatomical line diagrams, clinical photographs and ultrasound images — all of which provide definite visual enhancements to the accompanying text.

I recommend this book highly and extend my fullest congratulations to Dinesh on a job well done!

Professor Wilfred CG Peh, MD, MBBS, FRCPG, FRCPE, FRCR
Senior Consultant and Head, Department of Diagnostic Radiology,
Khoo Teck Puat Hospital
Clinical Professor, Yong Loo Lin School of Medicine,
National University of Singapore
Singapore, October 2020

Introduction

Ultrasound guided procedures have become an important skill for Musculoskeletal Radiologists, Sports Physicians, Extended Scope Physiotherapists and Radiographers. Indeed, it has superseded non-guided injection, due to its accuracy in delivery and the ability to avoid damage to surrounding tissue. However, ultrasound guided injections should ideally be undertaken by those who have had training and experience with musculoskeletal sonography, to ensure that correct structures and pathological changes are identified before an procedure is undertaken.

The purpose of this book is not to replace ultrasound education, hands-on training or guidance by an experienced mentor. Instead, it should serve as a quick reference or supplement for experienced users, so that they have an accompanying document when undertaking the procedures described and can therefore guide one's practice.

In addition to providing an outline for performing the procedures, there are tips on how to position the patient, guide the needle, and what sort of procedures can be undertaken in the different anatomical areas.

This book has been divided into the main anatomical areas, namely the upper limb (shoulder, elbow, hand and wrist), spine, lower limb (hip, knee and ankle and foot).

Within each anatomical area, the procedures are divided into those that focus on joints, tendons, bursae, nerves, ligaments, muscles and other structures. It is essential that the end user of this guide can identify these structures prior to undertaking injection, so as to differentiate what may or may not be treated. The key with ultrasound guided interventions is safety, due consideration and ensuring high standards of care when undertaking them, which includes practitioner skill.

As musculoskeletal medicine continues to expand, with better scanners and increasing demand, it is important that clinicians continue to look for ways to improve and expand their practice. Good luck with your development in musculoskeletal ultrasound and guided procedures. It is truly an exciting time to be involved in musculoskeletal medicine!

Ultrasound Guided Musculoskeletal Procedures in Sports Medicine. https://doi.org/10.1016/B978-0-323-91014-9.00009-0
Copyright © 2021 Elsevier Inc. All rights reserved.

Safety and consent

2.1 Safety

Safety of the patient and clinician is paramount throughout the injection process. For the patient, it reduces their risks from the procedure while for the clinician, it hopefully reduces the risk of incorrect injections, failure of the treatment and personal injuries such as from needlesticks. Taking a pragmatic, stepwise approach to your procedures, standardising care and planning ahead can help minimise undue risk to all stakeholders.

1. *Pre-scanning and preparation consultation.* It is useful to have the patient visit prior to the day of the injection, particularly if they have been referred by a colleague, so that they are aware of the position needed, the procedure itself and how to prepare for the procedure day. It also enables yourself, as the clinician, to plan for the procedure and decide the best approach.

2. *Confirmation of the patient and procedure.* This should be done on the day of the injection once the patient is in the treatment room and provides an opportunity for changes to be made based on current symptoms. If symptoms have resolved, it may prove unnecessary to proceed with the treatment.

3. *Final imaging and marking.* Repeating the imaging is helpful to remind the clinician about the anatomy, visualise the procedure and informs them if there have been any significant changes since the patient was last scanned. Marking also confirms the injection site for the clinical team and patient, helps minimise the risk of wrong site treatments and guides the clinician during the procedure.

4. *Time-out.* Prior to undertaking the cleaning, draping and preparation of medication, a formal "time-out" with confirmation of the patient's identity, clarification of the procedure and site, assures the patient and clinician before progressing.

5. *Sterile preparation of equipment.* Preparation of medication should only be undertaken once hands are cleaned and sterile gloves are worn, which should minimise contamination of the equipment. Thereafter, this equipment should only be handled by the clinician who has been suitably prepped.

6. *Cleaning and draping.* Preparing the skin with a fastidious approach is necessary to limit the development of post-injection infection and bleeding. Cleaning the transducer or using a drape helps limit this occurrence. Chlorhexidine gluconate

Ultrasound Guided Musculoskeletal Procedures in Sports Medicine. https://doi.org/10.1016/B978-0-323-91014-9.00001-6
Copyright © 2021 Elsevier Inc. All rights reserved.

0.05% solution is commonly used for the cleaning, but 10% povidone iodine can be used as an alternative.

7. *Post-injection care.* Ensuring that the skin is cleaned again after the procedure and applying an appropriate dressing should hopefully limit the development of a post-injection infection. Applying pressure can help with bruising or bleeding and immobilisation may be required if an intra-tendinous or ligamentous injection is undertaken to reduce the risk of tearing. Post-injection care should be individualised to the patient, taking into context their demographics, co-morbidities, occupation, social support and the procedure undertaken. If required, a written and suitably translated information sheet should be provided, with contact details for queries and emergencies.

2.2 Consent

Taking consent is essential prior to any procedure, to ensure the patient is aware of potential outcomes as well as risks. Although it may be taken prior to the procedure day, it is important to re-iterate what has been discussed before undertaking the procedure. Although consent is on an individual basis, it should broadly encompass:

1. *Clinical reasoning and confirmation of the type of procedure being undertaken.* This is particularly important if the patient has been referred from another clinician, who has been providing care up till now. It is essential that patients and clinicians alike understand why the procedure is being undertaken and that other, more conservative avenues of treatment have been explored prior to considering an injection. It is only with this understanding that a procedure should be contemplated and hence the reasoning must be re-iterated during the consent process.

2. *Expected outcomes and benefits.* What is the understanding of the patient and clinician in terms of a positive outcome? While the patient might hope that all their symptoms are resolved, the clinician may have a more measured expectation. Not addressing this, often leads to disappointment after the procedure and it is therefore important to arrive at a common set of anticipated outcomes. This is particularly relevant when managing degenerative disease, as the procedure may be undertaken to help with symptoms rather than the underlying disease process.

3. *Potential risks.* These are common to any type of injection and include infection, pain and bleeding at the site of the procedure. However, there may be other risks that need to be discussed based on the procedure being undertaken, the individual's co-morbidities or their pre-existing functional state.

4. *Potential side effects.* These may be due to the type of injection substrate being used (such as with steroids or prolotherapy), the procedure being undertaken (such as with intra-tendinous or intra-ligamentous techniques) and the patient's co-morbidities (such as diabetes). It is important to enhance understanding of

these factors by explaining how it can impact the patient and what they can do to minimise the possibilities.

5. *What to expect and do or not do post-procedure.* There may be certain things that the patient should or should not do post-injection and this might continue for a specified period of time. Indeed, it is important to help them understand the reason for this and therefore it hopefully optimises their outcome.

6. *Alternatives to the procedure.* Depending on what is available in the clinical setting, the clinician should have considered other options prior to undertaking the procedure. This might include engaging with other healthcare specialists such as physiotherapists, podiatrists, acupuncturists and even dieticians. Equally, if the procedure is being repeated and there has been limited success in the past, it might be prudent to discuss alternative interventions.

Injections

Various types of injections are available to the clinician and patient, but the decision to undertake a procedure on must be based on the clinical need and practitioner's experience. There must be a thorough clinical assessment, with evaluation of symptoms, and if needed further investigations (such as X-ray, CT or MRI) prior to discussing potential interventions. An additional consideration is the application of conservative treatments; engagement of the multi-disciplinary team is important prior to procedure and certainly may continue to be part of the overall paradigm afterwards.

Although not an exhaustive list, procedures commonly undertaken a typical Sports Medicine Centre include:

1. *Corticosteroid injections (CSIs).* A staple in many musculoskeletal clinics, CSIs can rapidly help settle pain and swelling in tissue, restoring function and allowing individuals to participate in their rehabilitation. Often combined with local anaesthetic, these injections are often described as having a diagnostic (from the anaesthetic) and therapeutic (from the steroid) element. Patients frequently question the duration of effect; generally, the first treatment is the most efficacious whereas subsequent injections may be less potent. However, this can vary on an individual basis. Commonly used injectable steroids in musculoskeletal medicine include triamcinolone acetonide, methylprednisolone acetate and dexamethasone. The choice depends on local availability and the site of treatment; for peri-radicular injections, particulate steroid preparations are generally not recommended due to the potential risk of vasospasm and a soluble form is recommended. For intra-articular injections, triamcinolone or methylprednisolone are typically used. *Potential side effects* to mention include a steroid flare, hypo-pigmentation, fat atrophy, a temporary increase in blood glucose, gastric irritation and vaginal spotting for female patients.

2. *Local anaesthetic (LA) injection.* Commonly combined with steroids in a CSI, these provide immediate analgesic effects to negate post-procedure pain as well as support clinicians from a diagnostic perspective. The former is important for patient comfort following procedures, while the latter can help identify whether the site injected is truly a pain-generator. There has been some debate on the use of intra-articular LA due to the risk of chondrotoxicity, however, for patient symptoms, a minimal should be considered. Commonly used LA includes lidocaine hydrochloride (1 or 2%), bupivacaine hydrochloride (0.25 or 0.5%) and ropivacaine hydrochloride. *Potential side effects* to mention include central

nervous system effects such as shivering, hypoventilation, respiratory arrest and convulsions with intrathecal administration. There is also the risk of cardiac arrest with intravascular delivery.

3. *Hyaluronic acid (HA) injections.* These have also become a staple in many musculoskeletal clinics particularly for mild to moderate osteoarthritis in a variety of joints. Preparations (high or low molecular weight) and delivery methods (single or multiple injections) may vary according to what is locally available; nevertheless HA injections tend to be useful in managing pain symptoms, particularly if combined with other conservative measures. *Potential side effects* to mention include a temporary flare of symptoms, swelling and tightness in the joint.

4. *Platelet-rich plasma (PRP) injections.* Various methods are available to prepare the PRP, with differing yields and concentrations of the platelets and plasma. Other systems also offer the ability to vary the concentration of white blood cells in the PRP. PRP injections have a wide variety of uses, but in this text, they will be used for intra-tendinous injections. There is also increasing evidence to suggest benefits in degenerative joint disease, in which case the steroid injection can be substituted with the PRP. *Potential side effects* to mention include a temporary flare of symptoms, swelling and tightness in the joint.

5. *Prolotherapy injections (Prolo).* Prolotherapy injections have been used for many years and for many different purposes. Equally, there are differing preparations of the prolotherapy that have been described. In the context of this book, the prolotherapy injections are used for ligament injuries, particularly where there is ongoing instability, but where surgical treatment may not be warranted. *Potential side effects* to mention include a temporary flare of symptoms, swelling and tightness in the surrounding tissue. Depending on the type of Prolo used, there might also be side effects from tissue irritation or a temporary increase in blood glucose (with dextrose).

6. *High volume injections (HVIs).* HVIs aim to create a plane in a particular tissue to help separate areas that might be otherwise tight or adherent. In doing so, the aim is to allow these areas to move more freely and enable the individual to undertake their rehabilitation. Sometimes a repeat injection can be required after the initial one should there be a plateau in progression of the clinical symptoms. It can be used as an alternative for areas where steroid should not be injected. *Potential side effects* to mention include a temporary flare of symptoms, swelling, stiffness or tightness in the tissue.

7. *Hydrodilatations.* Similar to HVIs, the purpose of the hydrodilatation is to place a large volume of fluid into a tissue space and release adhesions, thereby allowing the structure to move normally. Mainly used in the frozen shoulder, this must be followed by rehabilitation and sometimes repeat injections may be required. In the hydrodilatation, steroid is used to help with pain symptoms. *Potential side effects* to mention include a temporary flare of symptoms, swelling, stiffness and tightness in the joint.

Indications and contraindications for injections

4

When discussing injection treatments with patients, it is important to consider:

1. *What is the underlying pathology?* Not all pathologies may be amenable to treatment with an ultrasound guided intervention. Indeed, even for those that are concurrent treatment with rehabilitation and other conservative therapies often optimise the outcome. The advantage conferred by ultrasound guidance is that the clinician and patient can be certain that the area requiring treatment has been targeted and hence if the outcome is not positive, then perhaps another area is responsible for the symptoms.

2. *Is there need for further investigation?* It is important to be able to scan before undertaking injections. This is to avoid missing any unexpected pathologies when treating patients and consequently not treating pathologies inappropriately. While ultrasound is very helpful with superficial musculoskeletal pathologies, intra-articular and more complex pathologies might require further investigation such as with magnetic resonance imaging, to define the anatomy in greater detail. The disadvantage is that this is not a dynamic assessment and functional issues such as impingement or subluxation cannot be assessed.

3. *What other conservative treatments have been tried or should be considered?* Before injections are undertaken, it is important to consider non-invasive avenues for treatment, such as rehabilitation exercises, taping and other treatments with physiotherapists, orthoses with podiatrists and even treatments such as acupuncture. Engagement with this aspect of treatment is essential, as patients often require further rehabilitation following the intervention. Pre-procedure treatment can also help equip the patient with the skills necessary to optimise their outcome.

4. *Patient preferences?* A thorough discussion with the patient must be undertaken prior to undertaking the procedure. Even if, as a clinician, we feel that the procedure is an appropriate treatment, if the patient does not feel ready for it, they should not feel under pressure to go through with it. Ensure that your patient is ready for the treatment and then, if appropriate, provide this for them.

5. *How to manage expectations?* This is usually core to the consent-taking process, but it is important to ensure that the patient understands the following:
   a. The purpose of the treatment.
   b. What the treatment involves.

    c. What is required after the intervention as part of further treatment.
    d. The potential outcomes, including the possibility that it makes no difference to their symptoms or a repeat procedure is required.

4.1 Indications

1. *CSIs.* Providing a rapid analgesic and anti-inflammatory effect, CSIs can help with pain and swelling symptoms in multiple areas in the body. It is generally not advised to undertake CSI around weight-bearing tendons such as the Achilles or patella tendon due to the risk of rupture. Sometimes, this can be combined with other treatments, such as HA injections, particularly when there is acute pain or swelling.
2. *HA.* Recommended for pain associated with mild to moderate degenerative changes, the duration of effect will vary depending on the product and concurrent rehabilitation undertaken by the individual.
3. *PRP.* Generally advised for tendon injuries such as tears or chronic tendinopathy; there is also increasing evidence for the use of PRP in degenerative joint disease, particularly in weight-bearing structures, where it can help with pain symptoms.
4. *Prolo.* Prolo injections are typically used for partial ligament tears such as those around the ankle and knee, particularly where there are instability symptoms associated with it. It is important to immobilise the joint afterward to enable it to have the irritant and proliferative effect.
5. *HVIs.* Typically used where the tissue has become tight or adhesive, the HVI can help break up some of these areas such as in the subacromial bursa, with chronic bursitis, or even around ligaments following repeated sprains when scar tissue may have developed around it.
6. *Hydrodilatation.* These are typically used for significant tightness in a joint, such as with a frozen shoulder, but could potentially provide a similar effect elsewhere in the body. The volume will need to be adjusted according to the joint being treated.

4.2 Contraindications

These can be broadly divided into absolute or relative and while this is not an exhaustive list, some common ones include:

1. Absolute
    a. Allergy to the medication.
    b. Infection around injection site.
    c. Current systemic infection.

2. Relative
   a. Significant flare of the symptoms following previous treatments.
   b. Uncontrollable bleeding disorders or unchecked anti-coagulant therapy.
   c. Recent systemic infection.
   d. Immunosuppressant therapy.
   e. Severe aversion to needles.
   f. Previous failed injection treatments.
   g. Implants at or close to the injection site.

Equipment

While equipment used for injections depends on clinician preference, the procedure being undertaken, patient habitus and local availability, it is important to plan ahead and standardise where possible. By limiting variability, it enables clinical staff to evaluate outcomes more critically and identify areas where subtle yet significant changes can be made.

It is also important that clinical staff respect cleanliness as much as possible and organise their equipment prior to undertaking a procedure. Thus, it is essential to plan ahead, consider what is needed and what might be required should there be any challenges along the way.

A useful tip is laying out your equipment in the order that it will be used before undertaking the procedure. Again, by standardising this, it enables easier identification if anything has been omitted. It also enables clinicians to visualise the steps, from skin preparation, to guiding the needle itself.

In the following images, there are some suggestions on the equipment that is useful for standard CSI/HA, HVI, hydrodilatation, Prolo, PRP and nerve root/facet joint injections.

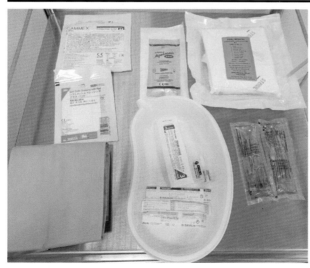

Set 1: CSI/HA

1. Sterile gloves
2. Sterile dressing pack
3. Sterile gel
4. Cleaning solution
5. Syringes
6. Needles
7. Dressing pack
8. Steroid, local anaesthetic

Ultrasound Guided Musculoskeletal Procedures in Sports Medicine. https://doi.org/10.1016/B978-0-323-91014-9.00010-7
Copyright © 2021 Elsevier Inc. All rights reserved.

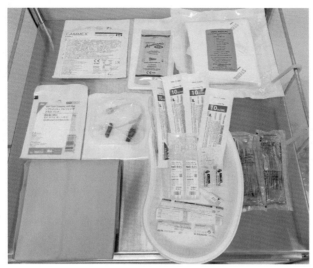

Set 2: HVI

1. Sterile gloves
2. Sterile dressing pack
3. Sterile gel
4. Cleaning solution
5. Syringes
6. Needles
7. Tubing
8. Dressing pack
9. Steroid, local anaesthetic, saline

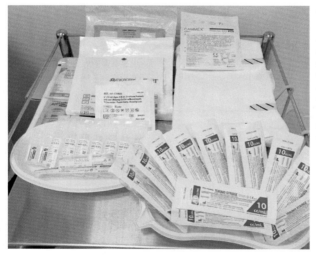

Set 3: Hydrodilatation

1. Sterile gloves
2. Sterile dressing pack
3. Sterile gel
4. Cleaning solution
5. Syringes
6. Needles
7. Tubing
8. Dressing pack
9. Steroid, local anaesthetic, saline

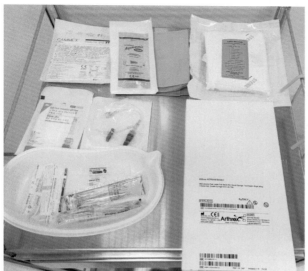

Set 4: PRP

1. Sterile gloves
2. Sterile dressing pack
3. Sterile gel
4. Cleaning solution
5. Syringes
6. Needles
7. Tubing
8. Dressing pack
9. PRP set, local anaesthetic

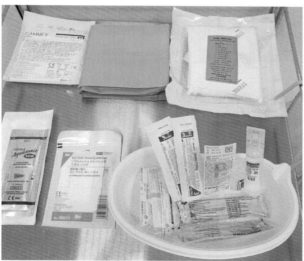

Set 5: Prolo

1. Sterile gloves
2. Sterile dressing pack
3. Sterile gel
4. Cleaning solution
5. Syringes
6. Needles
7. Tubing
8. Dressing pack
9. 50% Dextrose, local anaesthetic

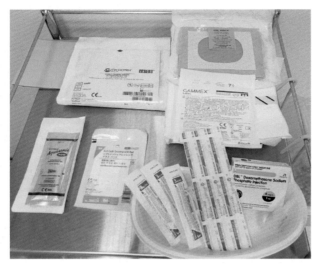

Set 6: Nerve root, facet and sacro-iliac joint injections

1. Sterile gloves
2. Sterile dressing pack
3. Sterile gel
4. Cleaning solution
5. Syringes
6. Needles
7. Dressing pack
8. Steroid, local anaesthetic

Post-injection protocols

6

It is important to provide patients with clear instructions following procedures to promote healing, optimise recovery and limit further injury. This may include:

1. Re-iterating the expected outcomes from the injection and potential side effects that could hinder their response to the treatment.
2. What to do if complications arise. It is useful to provide contact details for the clinic in case patients have any specific queries or concerns.
3. Instructions about offloading and a phased return to normal activities e.g. following PRP or Prolo injections. Initially, there may need to be a period of complete rest and offloading to allow the tissue to recover, followed by partial and eventually normal loading as tolerated.
4. Patients should be suitably trained in the use of appropriate equipment to help with offloading and immobilisation, e.g. crutches for lower limb, slings for shoulders, knee braces for patella tendons and boots for Achilles or plantar fascia treatments. It is useful to educate patients in the fitting, removal and maintenance of this equipment and also what to do if they have concerns about symptoms arising from its use.
5. Follow-up appointments at appropriate intervals to ensure that the tissue is healing and that post-procedure protocols are being followed. It provides patients and clinicians with a safety-net, should there be concerns about delayed recovery and further management.
6. The appropriate instigation and progression of rehabilitation exercises to enhance recovery and limit the recurrence of the injury.

Ultrasound Guided Musculoskeletal Procedures in Sports Medicine. https://doi.org/10.1016/B978-0-323-91014-9.00006-5
Copyright © 2021 Elsevier Inc. All rights reserved.

Abbreviations

1. *LA* — Local anaesthetic injection. The anaesthetic of choice depends on local availability and practice, but in the context of this pocketbook, it is 1% lidocaine.

2. *CSI* — Corticosteroid and LA injection. The steroid used depends on local availability, but in the context of this pocketbook it is 40mg/1 mL Triamcinolone for general injections and 4mg/1 mL Dexamethasone for spine injections. In situations where LA is omitted, this will be specified.

3. *PRP* — Platelet-rich plasma. The type of PRP remains hotly debated at the time of writing this book as is the extraction and preparation techniques. It will largely depend on the locally available system, but leucocyte-poor preparations are advocated for joint injections while leucocyte-rich is recommended for tendon procedures.

4. *HA* — Hyaluronic acid. Various types of visco-supplement are available on the market and which one is used will depend on clinician/institutional preference and local availability.

5. *Prolo* — Dextrose prolotherapy. This is made from a 1:1 dilution of 50% dextrose and 1% lidocaine.

6. *HVI* — High volume injections. This is usually achieved with the addition of 30—40 mls saline to the preparation.

7. *Hydro* — Hydrodilatation. The addition of up to 110 mls of saline, provided it is tolerated by the patient, provides the hydro-distention and dilatation effects of this procedure.

8. *mL* — millilitres.

9. *g* — gauge of the needle use.

10. *SAX* — Short-axis view. A cross-sectional view when imaging and undertaking the procedure.

11. *LAX* — Long-axis view. A longitudinal view when imaging and undertaking the procedure.

12. *IP* — In-plane approach. When undertaking the injection, the needle should be seen parallel to the transducer orientation and the length of it can be visualised.

13. *OOP* — Out-of-plane approach. When undertaking the injection, the needle is perpendicular to the transducer and will itself be seen in a cross-sectional perspective. It will be difficult to view the needle in its entirety in this orientation.

Ultrasound Guided Musculoskeletal Procedures in Sports Medicine. https://doi.org/10.1016/B978-0-323-91014-9.00003-X
Copyright © 2021 Elsevier Inc. All rights reserved.

Upper limb

Assessing and treating upper limb structures are commonly undertaken in musculo-skeletal medicine, particularly as these tend to be superficial and readily identified with ultrasound. These tend to become more amenable as you progress distally in the upper limb.

8.1. Shoulder

8.2. Elbow

8.3. Wrist

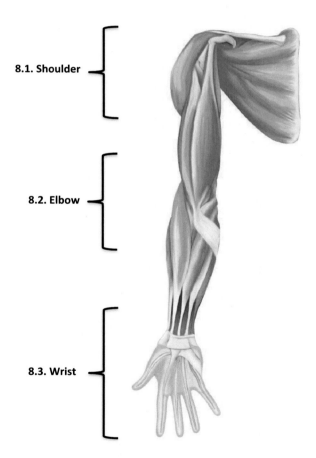

8.1 **Shoulder**

Superficial structures such as the rotator cuff, bursa and supra-scapular nerve can be evaluated using ultrasound. These and others, such as the acromioclavicular and glenohumeral joint, can be injected under ultrasound guidance. The key with shoulder injections is stability and limiting movement. If you have been trained in imaging and performing procedures in the seated position, it may be best to continue to do so; nevertheless, positioning the patient in a supine or prone position can optimise stability.

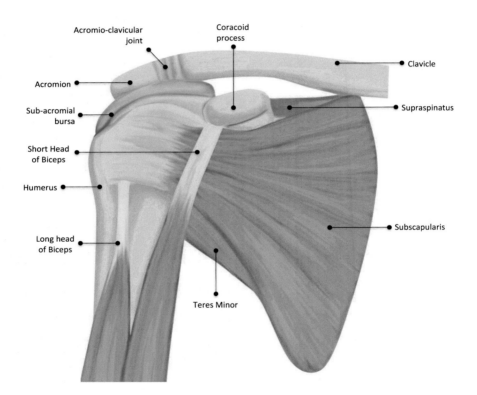

Joint injections

The glenohumeral (GHJ), acromioclavicular (ACJ) and sternoclavicular (SCJ) joints are commonly injected under ultrasound guidance for pain symptoms or limitation in movement. Positioning of the patient to aid the clinician can be an important consideration and this might need to be adapted to suit an individual basis.

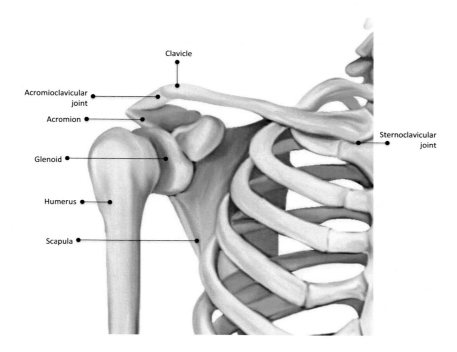

8.1.1 Glenohumeral joint

Patient position:	For the glenohumeral joint (GHJ), the patient can be positioned in a prone (Fig. 8.1.1A) or upright position (Fig. 8.1.1B). In the former, the arm hangs off the bed and gravity can be used to help open the GHJ. In the latter, the hand should rest on the opposite shoulder to optimise position and open the joint.
Identifying the anatomy:	In both patient positions, the GHJ can be traced back in LAX from the infraspinatus tendon insertion and along the curve of the humeral head (Fig. 8.1.1C—E). The posterior labrum may be identified and in degenerative shoulders, narrowing of the joint space or an effusion might be seen.
Injections performed:	CSI, PRP or HA injections into the shoulder for degenerative disease. With additional volume, a hydro can be performed for frozen shoulders.
Recommended transducer:	Curvilinear 3—5 mHz.
Equipment suggested:	*Equipment preparation:* Set 1 for CSI or HA injections. Set 3 for hydrodilatations. Set 4 for PRP injections. *Needle:* 2-inch 21- or 23-gauge needle. *Syringes:* 3 mL for cortisone injection or 10 mL(s) for a hydrodilatation. *Medication:* 40 mg triamcinolone (1 mL) and 1% lidocaine (5 mL). Add normal saline (110 mL) for hydrodilatations. Standard/available HA or PRP preparation.
Injection technique:	Using the humeral head as a guide, the needle can be inserted IP at approximately 45 degrees. With the bevel facing down, it can be guided into the posterior of the GHJ (Fig. 8.1.1F—H). As the joint capsule is breached, the patient may experience discomfort and if successfully placed, the injected solution should be seen flowing around the humeral head. A clampable extension tubing can help prevent fluid from escaping during a hydrodilatation. In situations where the needle is incorrectly placed, extravasation into the infraspinatus or surrounding tissue may be witnessed.

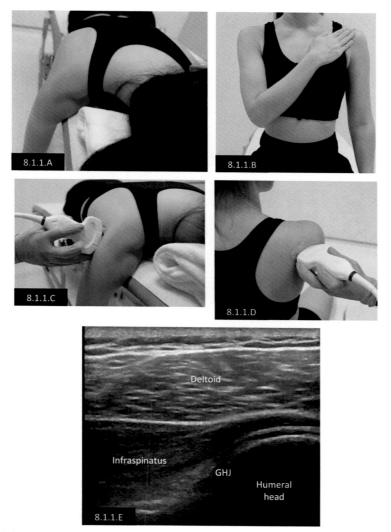

FIGURE 8.1.1 Injections to the GHJ.

The patient can be positioned in a prone or seated position with the arm positioned appropriately to open the joint. The injection can be undertaken IP, using the humeral head to guide the needle from lateral-to-medial. The needle tip should be seen within the GHJ.

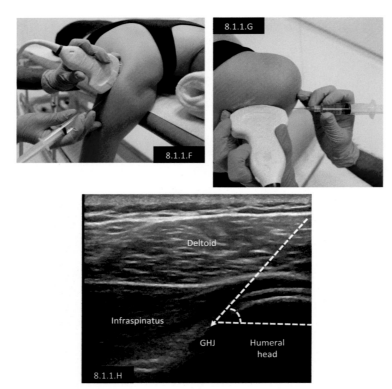

FIGURE 8.1.1 cont'd.

8.1.2 **Acromio-clavicular joint**

Patient position:	For the acromio-clavicular joint (ACJ), consider positioning the patient in a sitting (Fig. 8.1.2A) or supine position (Fig. 8.1.2B) with the arm by their side and the hand in a supinated position. The orientation may differ according to the side being injected and the clinician's hand dominance.
Identifying the anatomy:	The ACJ is readily identified by following the biceps tendon in the SAX proximally until the joint is visualised, or by palpating the clavicle and placing the transducer directly over the joint (Fig. 8.12C−E). Both enable the ACJ to be visualised in the LAX, where degenerative changes or synovial hypertrophy can be witnessed.
Injections performed:	CSI for pain, degenerative disease or synovitis. PRP or HA injections for degenerative disease.
Recommended transducer:	Linear 6−15 mHz. Hockey stick 8−18 mHz.
Equipment suggested:	*Equipment preparation:* Set 1 for CSI or HA injections. Set 4 for PRP. *Needle:* 1-inch 25- or 27-gauge needle. *Syringes:* 3 mL for CSI. *Medication:* 40 mg triamcinolone (1 mL) and 1% lidocaine (1 mL). Standard/available HA or PRP preparation.
Injection technique:	Maintaining the transducer in the LAX orientation, an IP technique can be used from a lateral approach to perform the injection with the needle at approximately 20 degrees (Fig. 8.1.2F−H). In situations where there is considerable degenerative change in the joint, an OOP approach can be used to access the joint by positioning the midline of the transducer over the joint and introducing the needle perpendicular to the skin (Fig. 8.1.2I−K).

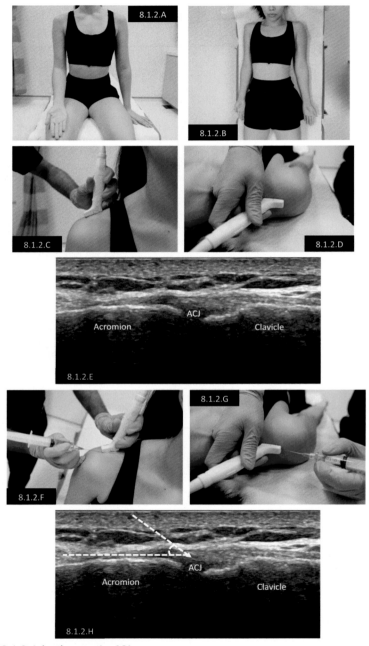

FIGURE 8.1.2 Injections to the ACJ.

The patient can be in a seated or supine position with the forearm also supine. The injection can be performed using an IP or OOP approach with the transducer in the LAX orientation. The needle tip should be seen within the ACJ.

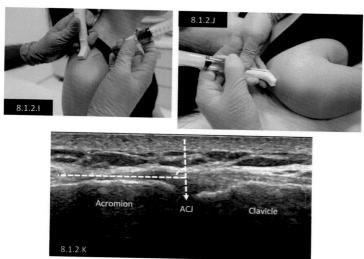

FIGURE 8.1.2 cont'd.

8.1.3 **Sternoclavicular joint**

Patient position:	For the sternoclavicular joint (SCJ), consider positioning the patient in a sitting (Fig. 8.1.3A) or supine position (Fig. 8.1.3B) with the arm by their side and the hand in a supinated position.
Identifying the anatomy:	The SCJ is readily identified by following the clavicle from lateral to medial or the sternum from an inferior to superior position. The joint can be identified in an LAX view as a gap between the clavicle and sternum (Fig. 8.1.3C—E).
Injections performed:	CSI for pain symptoms, degenerative disease or synovitis. PRP or HA injections for degenerative disease.
Recommended transducer:	Linear 6—15 mHz. Hockey stick 8—18 mHz.
Equipment suggested:	*Equipment preparation:* Set 1 for CSI or HA injections. Set 4 for PRP. *Needle:* 1-inch 25- or 27-gauge needle. *Syringes:* 3 mL for CSI. *Medication:* 40 mg triamcinolone (1 mL) and 1% lidocaine (1 mL). Standard/available HA or PRP preparation.
Injection technique:	Maintaining the transducer in the LAX, an IP technique can be used to perform the injection from a medial or lateral approach with the needle at 20 degrees and the bevel facing down (Fig. 8.1.3F—H). In situations where there is considerable degenerative change in the joint, an OOP approach can be used to access the joint by positioning the midline of the transducer over the joint and introducing the needle perpendicular to the skin (Fig. 8.1.3.I—K).

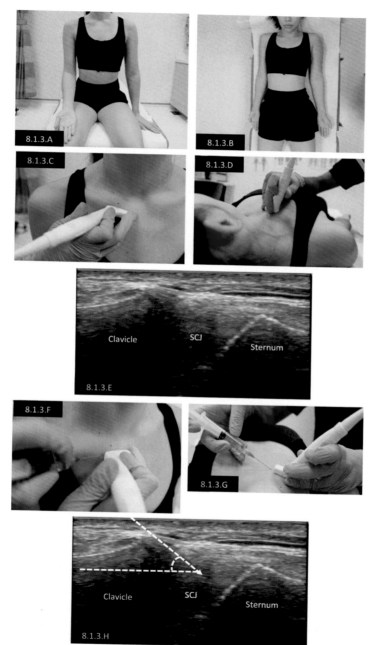

FIGURE 8.1.3 Injections to the SCJ.

The patient can be in a seated or supine position with the forearm also supine. The injection is undertaken using an IP (medial or lateral) or OOP approach with the transducer in the LAX orientation. The needle tip should be seen within the SCJ.

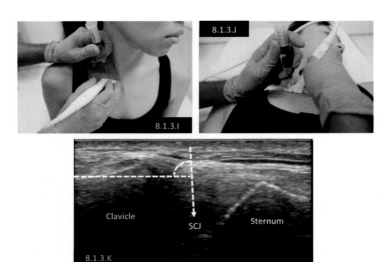

FIGURE 8.1.3 cont'd.

Tendons

With tendons around the shoulder readily visible using ultrasound, the commonest one requiring a direct procedure is the Long Head of the Biceps (LHBT) tendon when treating a tenosynovitis or tendinopathy. While pain in the others are more commonly treated through injection in to the Subacromial Subdeltoid Bursa, tears of the Supraspinatus can sometimes be amenable to a PRP injection.

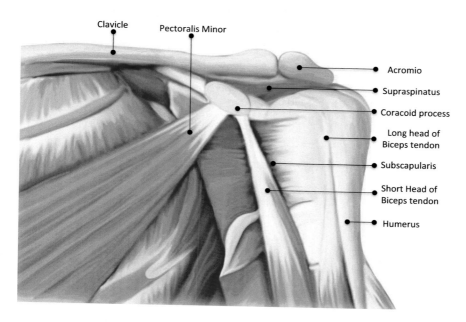

Clavicle

Pectoralis Minor

Acromio

Supraspinatus

Coracoid process

Long head of Biceps tendon

Subscapularis

Short Head of Biceps tendon

Humerus

8.1.4 **Long head biceps tendon**

Patient position:	For the long head biceps tendon (LHBT), the patient can be placed in a sitting (Fig. 8.1.4A) or supine position (Fig. 8.1.4B) with the hand supinated and the elbow flexed or extended respectively.
Identifying the anatomy:	Placing the transducer in transverse orientation the LHBT can be identified in the SAX within the bicipital groove. It should be traced distally to the pectoralis insertion, to check for integrity, prior to performing the injection (Fig. 8.1.4C–E). Fluid may be identified in the tendon sheath or tendinopathy might be noted. A further evaluation can be undertaken in an LAX view (Fig. 8.1.4F and G).
Injections performed:	CSI for tenosynovitis and pain. PRP for degenerative tendinopathy into the LHBT itself.
Recommended transducer:	Linear 6–15 mHz.
Equipment required:	*Equipment preparation:* Set 1 for CSI. Set 4 for PRP. *Needle:* 1.5- to 2-inch 25- or 27-gauge needle. *Syringes:* 3 mL for CSI. *Medication:* 20 mg triamcinolone (0.5 mL) and 1% lidocaine (1 mL). Standard/available PRP preparation.
Injection technique:	Using a lateral approach and the LHBT in an SAX orientation, the needle is introduced IP at approximately a 45-degree with the bevel facing down. Once the tip penetrates the LHBT sheath and the needle can be advanced until it makes contact with the bicipital groove (Fig. 8.1.4H–J). At this point the solution can be slowly injected and should be seen to flow around the LHBT, but if there is a resistance or flow into the tendon the needle tip must be readjusted. An alternative approach is in the LAX with the needle also in IP but this time at a 30–45 degrees (Fig. 8.1.4K and L). Once in the sheath, the fluid should be seen tracking along the length of the tendon. For PRP injections, it is important to identify the level of the injection and follow a similar technique, but in this situation the needle is inserted into the tendon itself before the solution is injected using a fenestration technique.

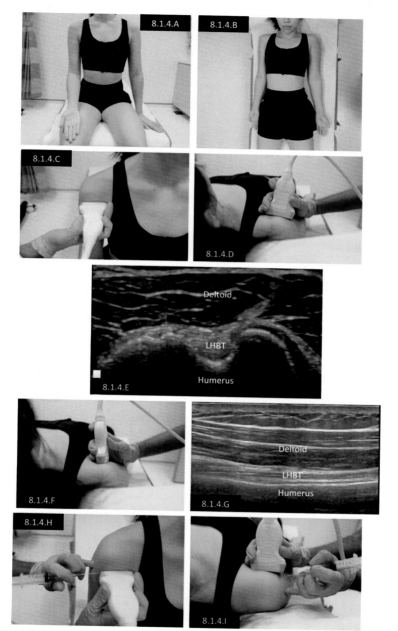

FIGURE 8.1.4 Injections to the LHBT.

The patient can be in a seated or supine position, with the hand also supine and the elbow flexed or extended. The injection is undertaken using an IP approach, with the transducer in a SAX or LAX orientation. The needle tip should be seen within the tendon and its sheath.

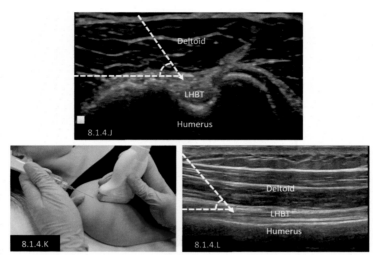

FIGURE 8.1.4 cont'd.

8.1.5 Supraspinatus

Patient position:	For the supraspinatus (SSP), consider positioning the patient in a supine but slightly side-rotated position with their hand in a back-pocket position (Fig. 8.1.5.A). A towel can be used to help the patient maintain this position. Alternatively, the patient can be sat upright, with the hand also in the back-pocket position (Fig. 8.1.5.B).
Identifying the anatomy:	Placing the transducer with one edge pointing towards the umbilicus (Fig. 8.1.5.C, D and E), the rotator interval can be identified with the long head of biceps tendon in SAX. Moving the transducer laterally will bring the SSP into view, also in an SAX orientation with the SASDB overlying. In this position SSP tears or tendinopathy can be identified, but must be subsequently confirmed in the LAX view.
Injections performed:	PRP for acute tears or degenerative tendinopathy with intrasubstance change.
Recommended transducer:	Linear 6—15mHz.
Equipment required:	*Equipment preparation:* Set 4 for PRP. *Needle:* 1.5-inch 23- or 25-gauge needle. Standard/available PRP preparation.
Injection technique:	Using a lateral approach with the transducer in the SAX, the needle is guided IP into the body of the SSP and into the tear (Fig. 8.1.5.F, G and H). With the bevel down, the injection is undertaken using a fenestration technique to suitably distribute the PRP. Care should be taken not to overly traumatise the intact tissue.

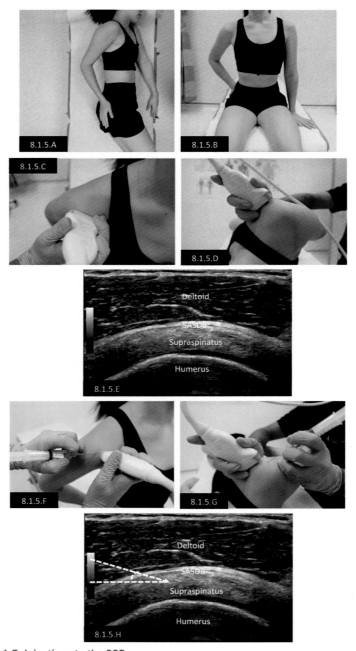

FIGURE 8.1.5 Injections to the SSP.

The patient can be in a seated or supine (side-lying) position with the hand on the back pocket. The injection can be undertaken using an IP approach, with the transducer in a SAX or LAX orientation. The needle tip should be seen within the SSP tendon itself.

Bursal injections

The subacromial subdeltoid bursa (SASDB) is a common site for shoulder injections when patients report pain with abduction or for impingement. It can also be helpful with generalised pain, limited movement or supraspinatus (SSP) tendinopathy. With chronic pain, there may be adhesions in the SASDB and a HVI can be useful to release these.

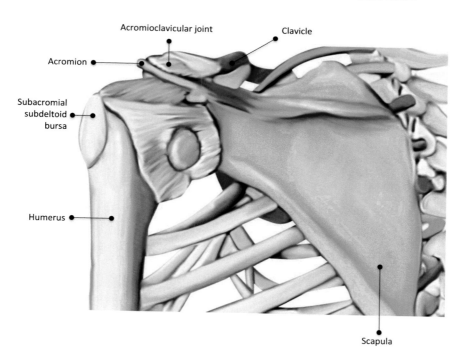

8.1.6 Subacromial subdeltoid bursa

Patient position:	For the subacromial subdeltoid bursa (SASDB), consider positioning the patient in a supine but slightly rotated position with the hand on the back pocket (Fig. 8.1.5A). A towel can be used to support the patient. Alternatively, the patient can be sitting, with the hand also in the back pocket position (Fig. 8.1.5B).
Identifying the anatomy:	Placing the transducer with one edge pointing toward the umbilicus (Fig. 8.1.5C–E), the rotator interval can be identified with the long head of biceps tendon in SAX. Moving the transducer laterally will be bring the SSP into view in the SAX with the SASDB overlying. In this position tears or tendinopathy can be identified and subsequently confirmed in an LAX. The bursa is identified and it can be assessed for thickening, bursitis or impingement.
Injections performed:	CSI for pain and impingement symptoms. HVI in chronic pain with adhesions.
Recommended transducer:	Linear 6–15 mHz.
Equipment required:	*Equipment preparation:* Set 1 for CSI. Set 2 for HVI. *Needle:* 1.5-inch 21- or 23-gauge needle. *Syringes:* 5 mL for CSI and 10 mls for HVI. *Medication:* 40 mg triamcinolone (1 mL) and 1% lidocaine (5–10 mL) for CSI and additional saline (20–30 mL) for HVI.
Injection technique:	Using a lateral approach with the transducer in the SAX, the needle is guided IP as horizontally as possible so that it can be visualised in its entirety (Fig. 8.1.5.F, G and H). With the bevel down, once the needle rests against the SSP a small amount of fluid can be injected to lift off the SASDB. Once separation is seen, the needle can be withdrawn slightly and the remainder of the solution can be injected into the SASDB. If there is pressure or it appears that the solution if being injected into the upper fibres of the tendon, the tip should be pulled back and repositioned. A clampable extension tubing can help prevent fluid from escaping during the HVI.

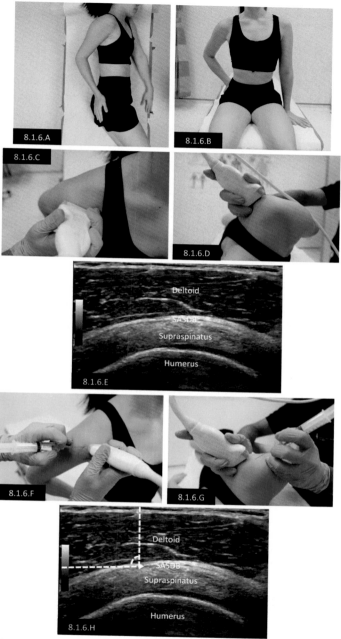

FIGURE 8.1.6 Injections to the SASDB.

The patient can be in a seated or side-lying position with the hand on the back pocket to add stability. The injection can be undertaken using an IP approach, with the transducer in a SAX or LAX orientation. The needle tip should be seen within the SASDB.

Nerve injections

Injection to the supra-scapular nerve (SSN) can be helpful during glenohumeral injections and for analgesia in chronic shoulder pain. A similar approach to glenohumeral joint injections can be undertaken but aiming more medial for the nerve in the supra-scapular notch.

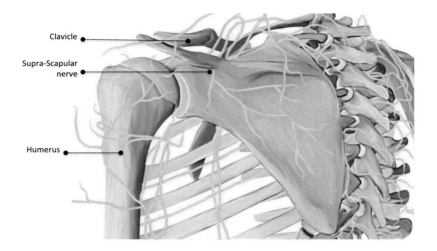

Clavicle

Supra-Scapular nerve

Humerus

8.1.7 **Supra-scapular nerve**

Patient position:	For the supra-scapular nerve (SSN), consider asking the patient to lie prone with the arm over the edge of the couch (Fig. 8.1.6A) or to be in a seated position with the hand resting on the opposite shoulder (Fig. 8.1.6B).
Identifying the anatomy:	Placing the transducer along with the spine of the scapula, this can be followed until the spinoglenoid notch is identified with SSN sitting within it in a SAX orientation (Fig. 8.1.6C—E). The vascular supply can be identified using the power Doppler and this should be avoided if possible.
Injections performed:	LA for diagnostic uncertainty or to relieve pain during shoulder procedures such as hydrodilatations. CSI for shoulder pain.
Recommended transducer:	Curvilinear 3—5 mHz.
Equipment required:	*Equipment preparation:* Set 1 for CSI or LA injection. *Needle:* 2- to 2.5-inch 23- or 25-gauge needle. *Syringes:* 3 mL for CSI or LA injections. *Medication:* 20 mg triamcinolone (0.5 mL) and 1% lidocaine (0.5 mL) for standard CSI and 1% lidocaine (1 mL) only for LA injections.
Injection technique:	Approaching from the lateral aspect, with the nerve in the SAX, the needle can be guided to the SSN within the spinoglenoid notch using an IP approach. The needle is aimed at approximately 45 degrees with the bevel pointing down to limit spilling of the solution. (Fig. 8.1.6F—H). Once the ligament is breached, and not blood flow is confirmed on aspiration, the solution can then be injected around the nerve.

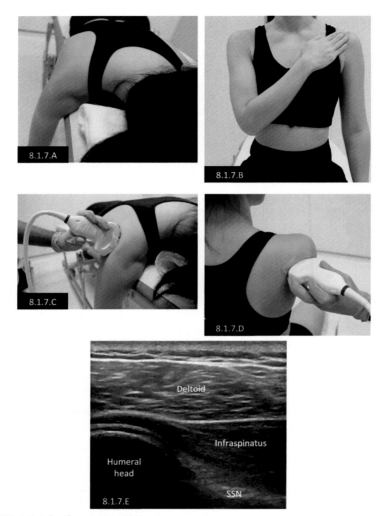

FIGURE 8.1.7 Injections to the SSN.

The patient can be positioned in a prone or seated position with the arm positioned suitably for stability. The needle can be guided into the supra-scapular notch, using an IP technique from lateral-to-medial, with the transducer in a LAX orientation. The needle tip should be seen beneath the ligament.

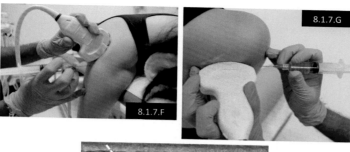

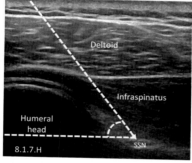

FIGURE 8.1.7 cont'd.

Notes
(Please use this area to reflect on your procedure and how you can build on these experiences).

8.2 Elbow

Superficial structures around the elbow can be readily identified and injected under ultrasound guidance with the patient either in a sitting or lying position. In both positions, the joint can be supported with a pillow or firm support to limit movement. By altering the elbow position, it can help bring the structure under treatment to a more superficial position.

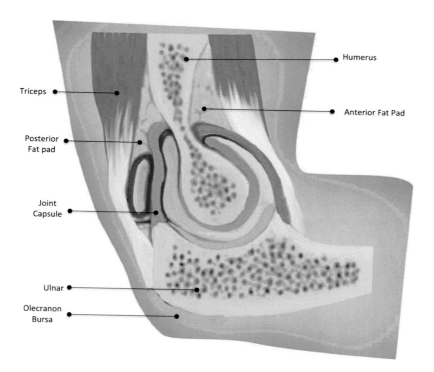

Humerus

Triceps

Anterior Fat Pad

Posterior
Fat pad

Joint
Capsule

Ulnar

Olecranon
Bursa

Joint injections

Although intra-articular aspects of the elbow joint cannot be assessed using ultrasound, superficial degenerative changes, such as osteophytes, can be seen. Guided treatments can be performed from the lateral aspect by injections into the Radio-Capitella joint (RCJ), for degenerative and pain-related symptoms, or the posterior joint line (PJL) aspect for degenerative changes or impingement extension. If the fat pad (PFP) is being caught, this can also be injected.

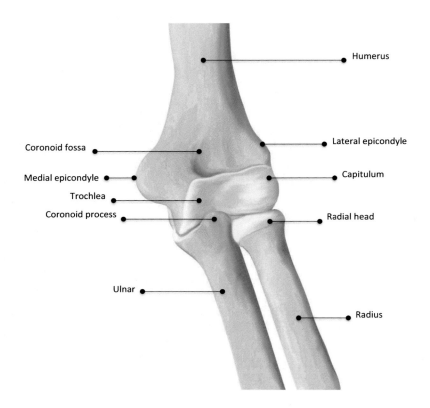

8.2.1 **Radio-capitella joint**

Patient position:	In the supine position, the patient can rest their hand against the body to expose the lateral elbow joint (Fig. 8.2.1A), while in the sitting position, the shoulder is internally rotated and the elbow is flexed (Fig. 8.2.1B). In both positions the hand is kept in a semi-supinated position.
Identifying the anatomy:	With the transducer along the length of the elbow, the radio-capitella joint (RCJ) can be seen in the LAX with the lateral epicondyle (LE) of the humerus and the radius in view. The common extensor tendon is seen inserting onto the LE and the radial collateral ligament (RCL) is situated beneath it (Fig. 8.2.1C–E).
Injections performed:	CSI for pain symptoms, degenerative disease and synovitis. PRP or HA injections for degenerative disease.
Recommended transducer:	Linear 6–15 mHz. Hockey stick 8–18 mHz.
Equipment required:	*Equipment preparation:* Set 1 for CSI or HA injections. Set 4 for PRP injections. *Needle:* 1- to 1.5-inch 23- or 25-gauge needle. *Syringes:* 3 mL for CSI. *Medication:* 40 mg triamcinolone (1 mL) and 1% lidocaine (1 mL) for standard CSI. Standard/available HA or PRP preparation.
Injection technique:	The injection can be performed using an IP approach, with the transducer in a LAX oreintation, guiding the needle from a distal to proximal position at approximately 45 degrees (Fig. 8.2.1F–H). It is important to ensure the bevel faces down into the joint to prevent extravasation. An OOP approach can also be used when there is considerable degenerative disease (Fig. 8.2.1I–K), with the transducer placed over the joint line and the needle entry perpendicular to the skin. In the latter, the needle can be seen passing vertically into the joint.

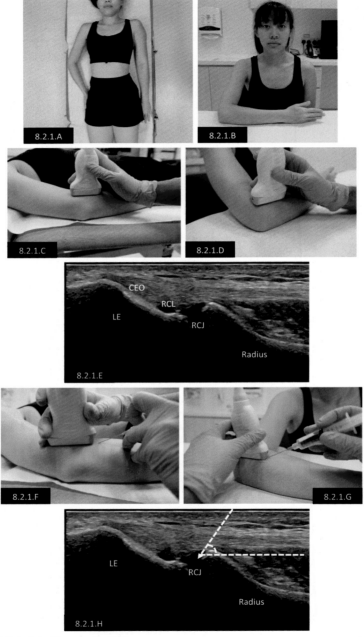

FIGURE 8.2.1 Injections to the RCJ.

The patient can be lying supine or seated with the elbow slightly flexed to open the joint and the hand placed against the body to add further stability. The injection can be undertaken using an IP approach, from a distal-to-proximal, with the transducer in the LAX orientation. The needle tip should be seen within the RCJ.

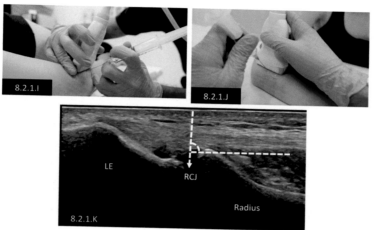

FIGURE 8.2.1 cont'd.

8.2.2 Posterior joint line and posterior fat pad

Patient position:	In the supine position, the patient can rest their palm flat on the surface of the treatment couch and with the elbow bent, they expose and open the posterior elbow (Fig. 8.2.2A). In the sitting position, a similar position can be used to access the area (Fig. 8.2.2B). Although both are somewhat awkward for the patient, by placing the palm flat, it enhances stability for the subsequent injection.
Identifying the anatomy:	In both patient positions, the triceps tendon can be followed in an LAX orientation to its insertion at the olecranon (Fig. 8.2.2C—E). Beneath this sits the posterior fat pad (PFP) and from here the posterior joint line (PJL) can be visualised. It can also be assessed in the SAX with the triceps seen in cross-section (Fig. 8.2.2F—H).
Injections performed:	CSI injections for fat pad impingement. PRP or HA injections for degenerative disease.
Recommended transducer:	Linear 6—15 mHz. Hockey stick 8—18 mHz.
Equipment required:	*Equipment preparation:* Set 1 for CSI or HA injections. Set 4 for PRP injections. *Needle:* 1- to 1.5-inch 25- or 27-gauge needle. *Syringes:* 3 mL for CSI. *Medication:* 20 mg triamcinolone (0.5 mL) and 1% lidocaine (1 mL) for standard CSI.
Injection technique:	The injection into the joint can be performed using an OOP approach with the transducer in the LAX and just adjacent to the medial border of the triceps tendon (Fig. 8.2.2I—K). Care must be taken not to take the needle into the substance of the TT. Alternatively, the transducer can be placed in the SAX over the tendon and the needle is introduced using an IP approach at approximately 45 degrees (Fig. 8.2.2L—N). The former is more suited to a posterior joint and the latter for a fat pad injection. In both situations it is important to be aware of the ulnar and radial nerve positions to avoid potential injury.

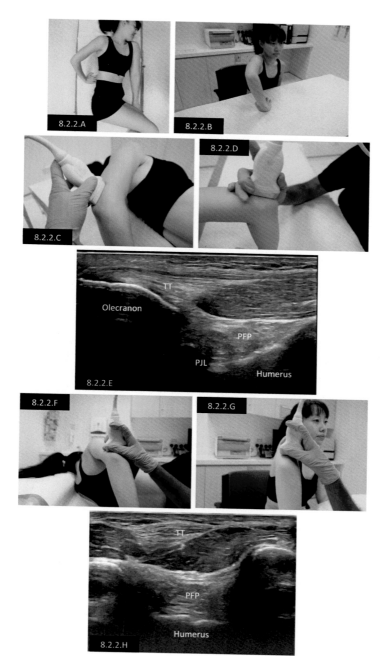

FIGURE 8.2.2 Injections to the PJL and PFP.

In both the lying and seated positions, the patient can place their hand flat on the treatment couch, with the posterior elbow exposed, to add further stability during the procedure. The injection can be undertaken using an IP or OOP approach with the transducer in a LAX or SAX orientation. The needle tip should be seen within the joint or PFP.

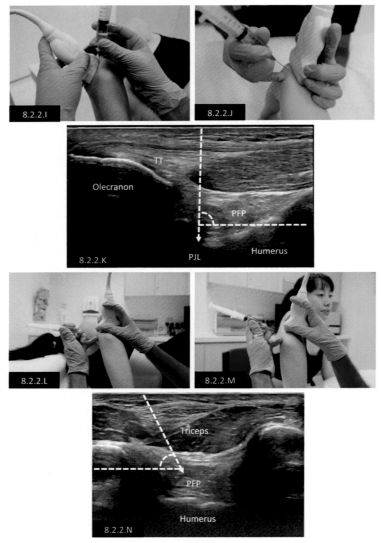

FIGURE 8.2.2 cont'd.

Tendon injections

Tendons commonly causing pain or functional symptoms around the elbow include the common extensor origin (CEO) or common flexor origin (CFO) and less often the triceps tendon (TT).

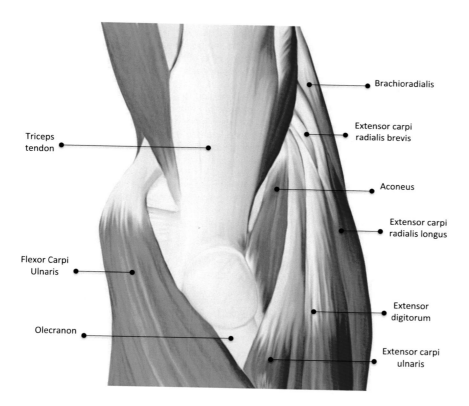

8.2.3 Common extensor origin

Patient position:	In the supine position, the patient can rest their arm flat on the treatment couch, with the elbow bent and the hand in a semi-supinated position (Fig. 8.2.3A). Resting the hand against the body adds further stability. A similar position can be taken in the seated position with the arm resting atop of the treatment couch (F 8.2.3B). Both will enable access to the common extensor origin (CEO).
Identifying the anatomy:	In the LAX orientation, the lateral epicondyle (LE) of the humerus and radial head can be identified, with the common extensor tendon inserting onto the LE and the radial collateral ligament beneath (Fig. 8.2.3C—E). It should also be visualised in an SAX view and also care should be taken to look out for the radial nerve and its branches.
Injections performed:	CSI for pain (although this is not generally advocated). HVI (tendon stripping) for tendinopathy with neovascularity. PRP for degenerative tendinopathy with intrasubstance tears.
Recommended transducer:	Linear 6—15 mHz. Hockey stick 8—18 mHz.
Equipment required:	*Equipment preparation:* Set 1 for CSI. Set 2 for HVI. Set 4 for PRP. *Needle:* 1- to 1.5-inch 25- or 27-gauge needle. *Syringes:* 3 mL for CSI, 10 mL for HVI. *Medication:* 20 mg triamcinolone (0.5 mL) and 1% lidocaine (1 mL) for standard CSI. Normal saline (20 —30 mL) can be added for HVI.
Injection technique:	The injection can be performed using an IP approach (Fig. 8.2.3F—H), guiding the needle from a distal to proximal direction at approximately 15 degrees. It is important to ensure the bevel faces down and it can be brought to rest upon the tendon before injecting a small volume to separate this from the overlying tissue. Once this separation is noted, the needle can be withdrawn and the HVI (tendon stripping) can be undertaken or the CSI can be delivered to this space. In situations where a PRP injection is undertaken for the tendon, the needle is introduced into the tendon and a fenestration technique is used to deliver it into the structure. Care must be taken to avoid injury to the Radial nerve.

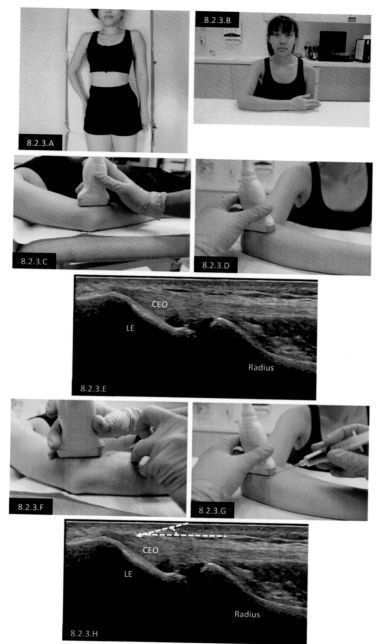

FIGURE 8.2.3 Injections to the CEO.

The patient can be lying supine or seated with the elbow slightly flexed to open the joint and the hand placed against the body to add further stability. The injection can be undertaken using an IP approach, from a distal-to-proximal, with the transducer in the LAX orientation. The needle tip should be seen above or within the CEO.

8.2.4 Common flexor origin

Patient position:	In the sitting position, the arm can be placed in a supinated and externally rotated position to expose the common flexor origin (CFO) (Fig. 8.2.4A). Supine, the patient can have the arm supinated and slightly abducted (Fig. 8.2.4B). In the prone position, the patient can rest their arm flat on their back with the elbow flexed to 90 degrees and the hand in a semi-pronated position (Fig. 8.2.4C). The latter tends to be a more stable position to perform a procedure.
Identifying the anatomy:	With the transducer in the LAX orientation, the medial epicondyle (ME) of the humerus and ulnar can be identified, with the CFO inserting onto the ME. The ulnar collateral ligament is located beneath (Fig. 8.2.4D—G). This can be seen in all three patient positions. An SAX view can help to visualise the ulnar nerve prior to undertaking the injection.
Injections performed:	CSI for pain (although this is not generally advocated). HVI (tendon stripping) for neovascularity. PRP for degenerative tendinopathy with intrasubstance tears.
Recommended transducer:	Linear 6—15 mHz. Hockey stick 8—18 mHz.
Equipment required:	*Equipment preparation:* Set 1 for CSI. Set 2 for HVI. Set 4 for PRP. *Needle:* 1- to 1.5-inch 25- or 27-gauge needle. *Syringes:* 3 mL for CSI, 10 mL(s) for HVI. *Medication:* 20 mg triamcinolone (0.5 mL) and 1% lidocaine (1 mL) for standard CSI. Normal saline (20 —30 mL) can be added for HVI.
Injection technique:	The injection can be performed using an IP approach (Fig. 8.2.4H—K), guiding the needle from a distal to proximal direction at an approximately 15 degrees. It is important to ensure the bevel faces down and it can be brought to rest upon the tendon before injecting a small volume to separate this from the overlying tissue. Once this separation is noted, the needle can be withdrawn and the HVI (tendon stripping) is undertaken or the CSI can be delivered to this space. If a PRP injection is being undertaken for the tendon, the needle is passed into the tendon and a fenestration technique is used to distribute the solution within the structure.

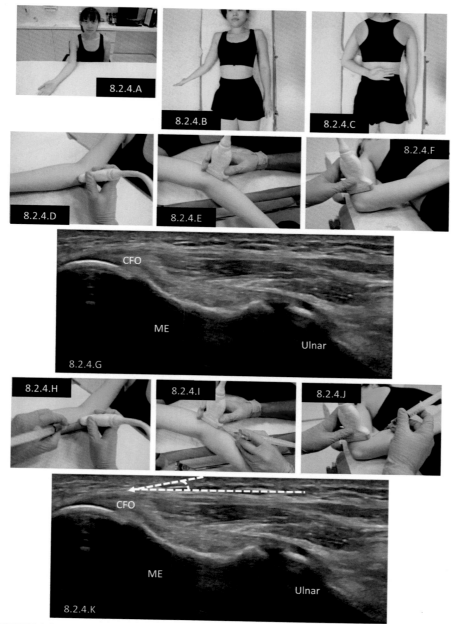

FIGURE 8.2.4 Injections to the CFO.

The patient can be seated or lying supine with elbow slightly flexed and shoulder externally rotated, or prone with the shoulder internally rotated and the arm on the back. The injection can be undertaken using an IP approach, from a distal-to-proximal, with the transducer in the LAX orientation. The needle tip should be seen above or within the CFO.

8.2.5 **Triceps Tendon**

Patient position:	In the supine position, the patient can rest their palm flat on the surface of the treatment couch and keep the elbow bent. This position exposes the posterior elbow and the triceps tendon (TT) (Fig. 8.2.5.A). In the sitting position, a similar position can be used to access the area (Fig. 8.2.5.B). Although both are somewhat awkward for the patient, but by placing the palm flat, it enhances stability for the subsequent injection.
Identifying the anatomy:	In both patient positions, the TT can be followed in an LAX orientation to its insertion at the Olecranon (Fig. 8.2.5.C, D and E). It can also be assessed in the SAX by rotating the transducer 90° (Fig. 8.2.5.F, G and H).
Injections performed:	PRP tendon pathologies such as tendinopathy with degenerative changes.
Recommended transducer:	Linear 6—15mHz. Hockey stick 8—18mHz.
Equipment required:	*Equipment preparation:* Set 4 for PRP injections. *Needle:* 1-1.5-inch 25- or 27-gauge needle. Standard/available PRP preparation.
Injection technique:	The injection into the TT can be performed using an IP approach with the transducer in the LAX (Fig. 8.2.5.F, G and H) or SAX (Fig. 8.2.5.L, M and N). In both, the needle is introduced at approximately 20° and the solution is injected using a fenestration technique. In both orientations it is important to be aware of the ulnar and radial nerve positions to avoid potential injury.

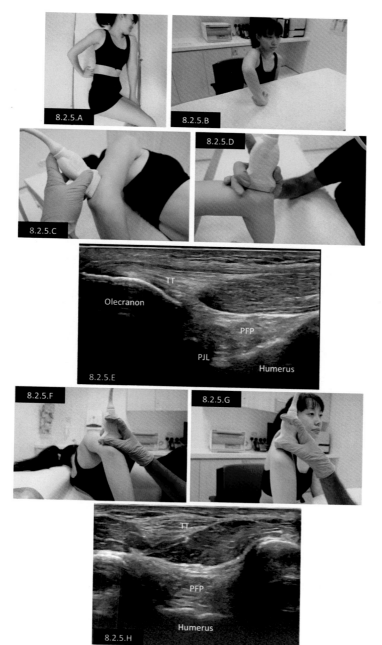

FIGURE 8.2.5 Injections to the TT.

In both the side-lying and seated positions, the patient can place their hand flat on the treatment couch, with the posterior elbow exposed, to add further stability during the procedure. The injection can be undertaken using an IP or OOP approach with the transducer in a LAX or SAX orientation. The needle tip should be seen within the TT.

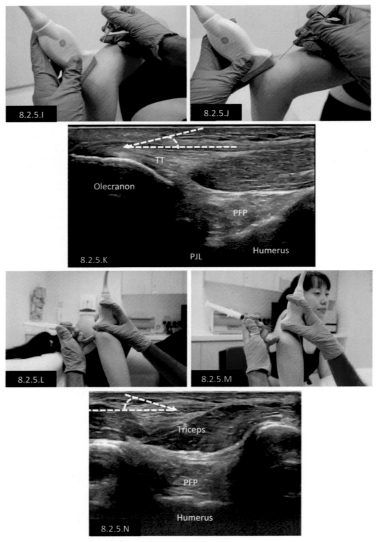

FIGURE 8.2.5 cont'd.

Ligament injection

The two most commonly injected ligaments in the elbow include the laterally placed radial collateral (RCL) or the medially situated ulnar collateral (UCL). These may cause pain and stability issues around the elbow, particularly with repeated throwing activities or a sudden trauma.

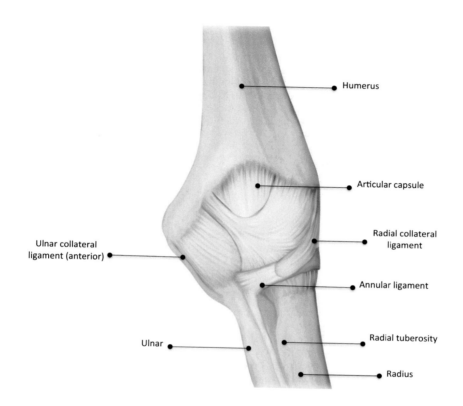

Humerus

Articular capsule

Radial collateral ligament

Ulnar collateral ligament (anterior)

Annular ligament

Ulnar

Radial tuberosity

Radius

8.2.6 Radial collateral ligament

Patient position:	In the supine position, the patient can rest their arm flat on the treatment couch, with the elbow bent and the hand in a semi-supinated position (Fig. 8.2.6A). Resting the palm against the body adds further stability. A similar forearm position can be taken in the seated position with the arm resting atop of the treatment couch (Fig. 8.2.6B). Both will enable access to the radial collateral ligament (RCL).
Identifying the anatomy:	With the transducer in a LAX orientation, the lateral epicondyle (LE) of the humerus and radial head can be identified, with the RCL beneath the common extensor origin (Fig. 8.2.6C—E). It can also be visualised in an SAX view and care should be taken to look out for the radial nerve and its branches.
Injections performed:	CSI for pain symptoms. Prolo injections for ligament tears.
Recommended transducer:	Linear 6—15 mHz. Hockey stick 8—18 mHz.
Equipment required:	*Equipment preparation:* Set 1 for CSI. Set 5 for Prolo injections. *Needle:* 1- to 1.5-inch 25- or 27-gauge needle. *Syringes:* 3 mL for CSI and Prolo. *Medication:* 20 mg triamcinolone (0.5 mL) and 1% lidocaine (1 mL) for standard CSI. For Prolo, 2 mL of a 50:50 mixture of 50% dextrose and 1% lidocaine can be used.
Injection technique:	The injection can be performed using an IP approach (Fig. 8.2.6F—H), guiding the needle from a distal to proximal direction at approximately 45 degrees. It is important to ensure the bevel faces down and it can be brought to rest upon RCL before injecting a small volume to separate this from the overlying tissue. Once this separation is noted, the needle can be withdrawn and the CSI can be delivered to this space. In situations where Prolo injections are undertaken, the needle is passed into the RCL and a fenestration technique can be used to deliver the solution.

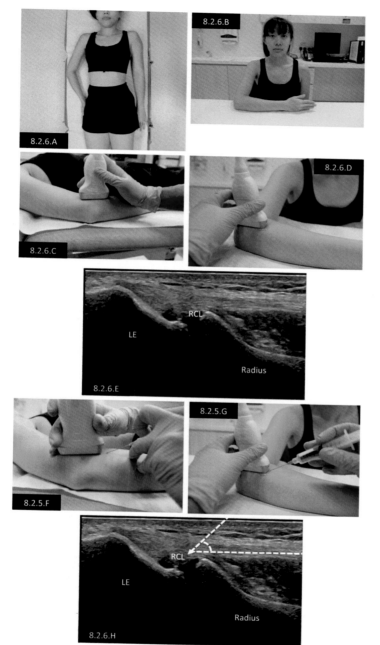

FIGURE 8.2.6 Injections to the RCL.

The patient can be lying supine or seated with the elbow slightly flexed. Placing the hand against the body further stabilises the elbow. The injection can be undertaken IP, from a distal-to-proximal approach, with transducer in a LAX orientation. The needle tip should be seen within the RCL.

8.2.7 Ulnar collateral ligament

Patient position:	In the sitting position, the arm can be placed in a supinated and externally rotated position to expose the ulnar collateral ligament (UCL) (Fig. 8.2.7A). In supine, the patient can have the arm supinated and slightly abducted (Fig. 8.2.7B). In the prone position, the patient can rest their arm flat on their back with the elbow flexed to 90 degrees and the hand in a supinated position (Fig. 8.2.7C). The latter tends to be a more stable position to perform a procedure.
Identifying the anatomy:	In the LAX view the medial epicondyle (ME) of the humerus and ulnar can be identified, with the UCL beneath the common flexor origin (CFO) (Fig. 8.2.7D–G). This can be seen in all three patient positions. An SAX view can help to visualise the ligament further.
Injections performed:	CSI for pain symptoms. Prolo injections for ligament tears.
Recommended transducer:	Linear 6–15 mHz. Hockey stick 8–18 mHz.
Equipment required:	*Equipment preparation:* Set 1 for CSI. Set 2 for HVI. Set 5 for Prolo. *Needle:* 1- to 1.5-inch 25- or 27-gauge needle. *Syringes:* 3 mL for CSI or Prolo. *Medication:* 20 mg triamcinolone (0.5 mL) and 1% lidocaine (1 mL) for standard CSI. For Prolo, 2 mL of a 50:50 mixture of 50% dextrose and 1% lidocaine can be used.
Injection technique:	The injection can be performed using an IP approach (Fig. 8.2.7H–K), guiding the needle from a distal to proximal direction at an approximately 45 degrees. It is important to ensure the bevel faces down and it can be brought to rest upon the UCL before injecting a small volume to separate this from the overlying CFO. Once this separation is noted, the needle can be withdrawn and the CSI can be delivered to this space. In situations where Prolo injections are undertaken, the needle is passed into the UCL and a fenestration technique can be used to deliver the solution.

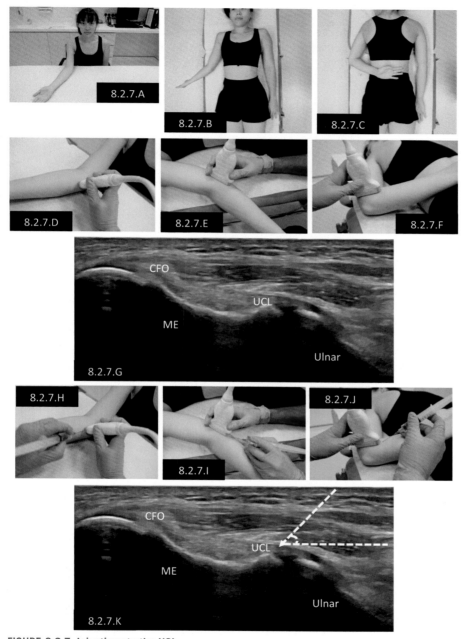

FIGURE 8.2.7 Injections to the UCL.

The patient can be lying supine or seated with the elbow slightly flexed and the shoulder slightly externally rotated, or in a prone position with the shoulder internally rotated and the forearm resting on the back. The injection can be undertaken using an IP approach, from a distal-to-proximal, with the transducer in a LAX orientation. The needle tip should be seen within the UCL.

Bursa

Two areas for a bursa are around the distal biceps tendon (DBTB) and olecranon (OB), usually triggered off by constant pressure on the olecranon or overload of the distal biceps tendon.

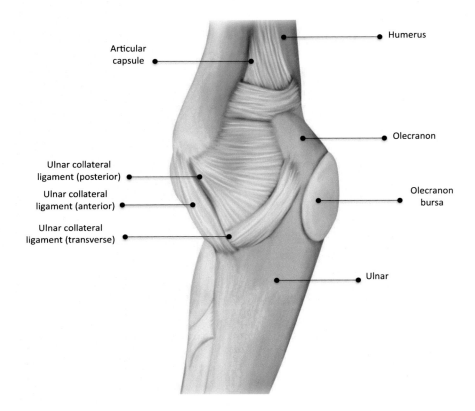

Humerus

Articular capsule

Olecranon

Ulnar collateral ligament (posterior)

Ulnar collateral ligament (anterior)

Olecranon bursa

Ulnar collateral ligament (transverse)

Ulnar

8.2.8 Distal biceps tendon bursa

Patient position:	In supine, the patient can position their arm with the elbow flexed to 90 degrees with the forearm pronated (Fig. 8.2.8A). A similar position can be used in with the patient sitting (Fig. 8.2.8B).
Identifying the anatomy:	With the elbow flexed and forearm pronated, the transducer is placed in an SAX between the ulnar and radius. The distal biceps tendon can be seen wrapping around the radius and the distal biceps tendon bursa (DBTB) will be seen sitting above this (Fig. 8.2.8.C—E).
Injections performed:	CSI for bursitis or impingement symptoms.
Recommended transducer:	Linear 6—15 mHz.
Equipment required:	Hockey stick 8—18 mHz.
	Equipment preparation: Set 1 for CSI.
	Needle: 1- to 1.5-inch 25- or 27-gauge needle.
	Syringes: 3 mL for CSI.
	Medication: 40 mg triamcinolone (1 mL) and 1% lidocaine (1 mL) for standard CSI.
Injection technique:	The injection can be performed using an IP approach (Fig. 8.2.8F—H), guiding the needle at an approximately 40 degrees. It is important to ensure the bevel faces down and it can be brought to rest upon the tendon before injecting a small volume to separate the bursa from the surrounding tissue. Once this separation is noted, the needle can be repositioned within the bursa and the solution should be seen flowing freely.

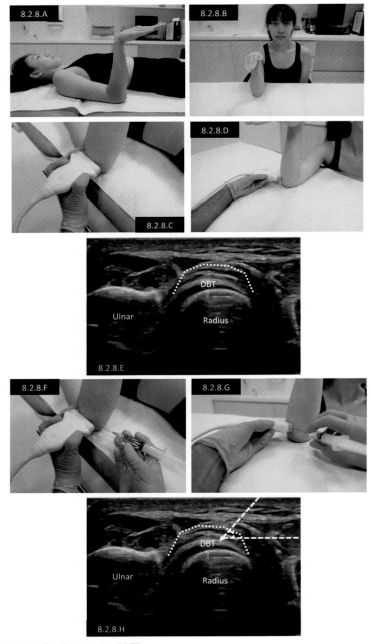

FIGURE 8.2.8 Injections to the DBTB.

The patient can be lying supine or seated with the elbow flexed to 90° and the forearm pronated. The injection can be undertaken using an IP approach, from medial-to-lateral, with the transducer in a LAX orientation. The needle tip should be seen over the distal biceps tendon within the DBTB.

8.2.9 Olecranon bursa

Patient position:	In the supine position, the patient can rest their extended and slightly internally rotated arm on the surface of the treatment couch to expose the posterior elbow (Fig. 8.2.9A). In the sitting position, a similar position can be used to access the area (Fig. 8.2.9B).
Identifying the anatomy:	In both patient positions, the triceps tendon can be followed in the LAX to its insertion at the olecranon and following this over the edge of the elbow, the olecranon bursa (OB) can be seen (Fig. 8.2.9C–E). The OB can also be assessed in the SAX view by turning the transducer 90 degrees (Fig. 8.2.9F–H).
Injections performed:	CSI injections following aspiration of the OB can be undertaken.
Recommended transducer:	Linear 6–15 mHz. Hockey stick 8–18 mHz.
Equipment required:	*Equipment preparation:* Set 1 for CSI injections and aspirations. *Needle:* 1- to 1.5-inch 25- or 27-gauge needle. *Syringes:* 3 mL for CSI and 10 mL(s) for aspiration. *Medication:* 20 mg triamcinolone (0.5 mL) and 1% lidocaine (1 mL) for standard CSI injections.
Injection technique:	The injection into the OB can be performed using an IP technique in an LAX (Fig. 8.2.9I–K) or SAX (Fig. 8.2.9L–N) approach. The needle is introduced at approximately 20 degrees with the bevel pointing down. Once the aspiration is undertaken, the syringe can be switched to the one containing the CSI, thereby minimising the need for repeat punctures. If the bursal has completely settled following aspiration, a small amount should be injected to reopen the space before the needle is repositioned and the remainder injected.

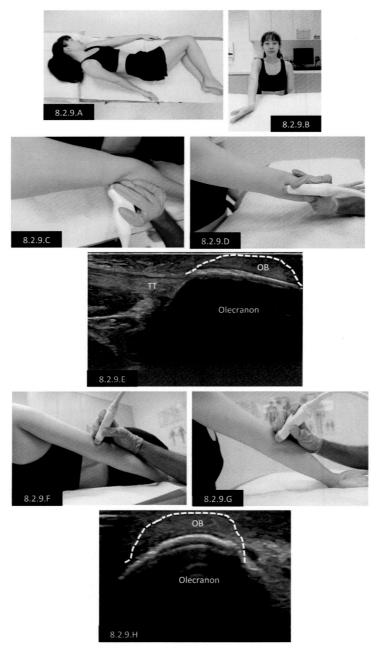

FIGURE 8.2.9 Injections to the OB.

The patient can be side-lying or seated with the elbow extended flexed and the shoulder slightly internally rotated. The injection can be undertaken using an IP approach, with the transducer in a SAX or LAX orientation. The needle tip should be seen within the OB.

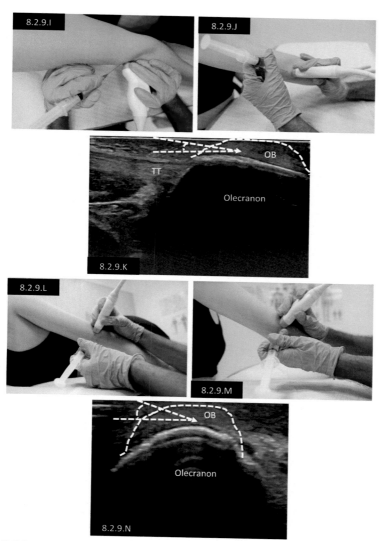

FIGURE 8.2.9 cont'd.

Nerve injections

The ulnar nerve (UN) is the commonest nerve requiring injection around the elbow as it passes through the cubital tunnel. Symptom can arise from subluxation or prolonged compression.

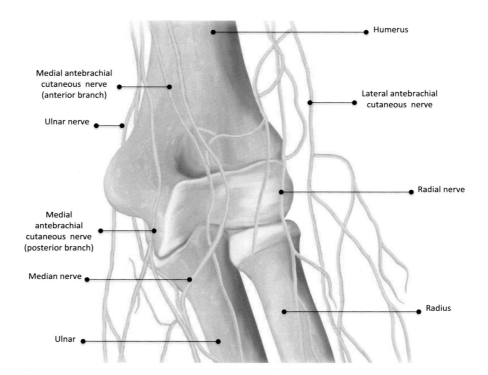

8.2.10 **Ulnar nerve**

Patient position:	In the supine position, the patient can rest their palm flat on the surface of the treatment couch and with the elbow bent, they expose the posterior-medial elbow (Fig. 8.2.10A). In the sitting position, a similar position can be used to access the area (Fig. 8.2.10B). Although both are somewhat awkward for the patient, but by placing the palm flat, it enhances stability for the subsequent injection.
Identifying the anatomy:	Placing the transducer across the cubital tunnel, the ulnar nerve (UN) can be identified in a SAX view within the cubital tunnel (Fig. 8.2.10C—E), sitting adjacent to the bone. It should be traced proximally and distally for integrity and also reviewed in LAX for swelling or thickening. Subluxation over the medial epicondyle can be assessed by passively extending the elbow.
Injections performed:	CSI injections for neuropathy from subluxation or entrapment.
Recommended transducer:	Hockey stick 8—18 mHz.
Equipment required:	*Equipment preparation:* Set 1 for CSI. *Needle:* 1- to 1.5-inch 25- or 27-gauge needle. *Syringes:* 3 mL for CSI. *Medication:* 20 mg triamcinolone (0.5 mL) and 1% lidocaine (0.5 mL) for standard CSI.
Injection technique:	The injection can be performed in the SAX view using an IP technique with the needle angled at approximately 20—30 degrees (Fig. 8.2.10F—H). With the bevel down, the solution is injected into the cubital tunnel space once the ligament is breached, with care to avoid injecting the nerve itself. Alternatively, an OOP technique can be used with the transducer still in the SAX position (Fig. 8.2.10I—K). The needle is brought into the cubital tunnel above the UN and the solution is delivered.

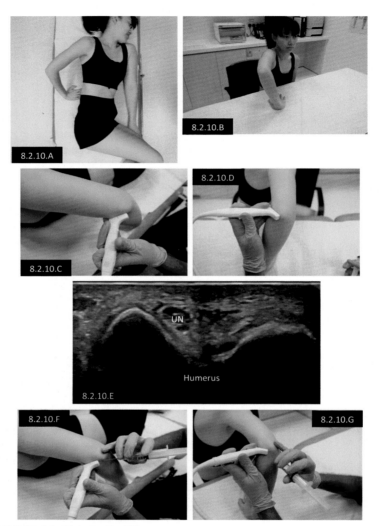

FIGURE 8.2.10 Injections to the UN.

In both the side-lying and seated positions, the patient can place their hand flat on the treatment couch, with the posterior elbow exposed, to add further stability when performing the procedure. The injection can be undertaken using an IP or OOP, with the transducer in a SAX orientation. The needle tip should be seen within Guyon's canal, adjacent to the UN.

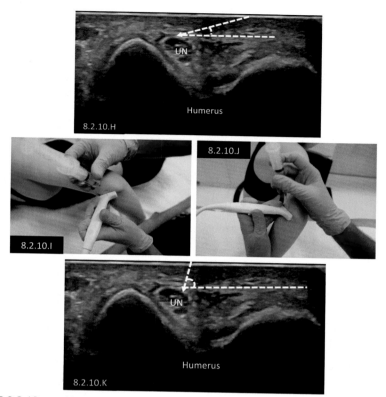

FIGURE 8.2.10 cont'd.

Notes
(Please use this area to reflect on your procedure and how you can build on these experiences).

8.3 **Wrist and hand**

Most structures around the wrist and hand can be treated under ultrasound guidance due to their superficial nature. Intra-articular pathologies can be injected using ultrasound guidance but cannot be investigated. The patient can be either in a sitting or lying position for wrist procedures with the joint supported by a pillow or firm support to limit movement.

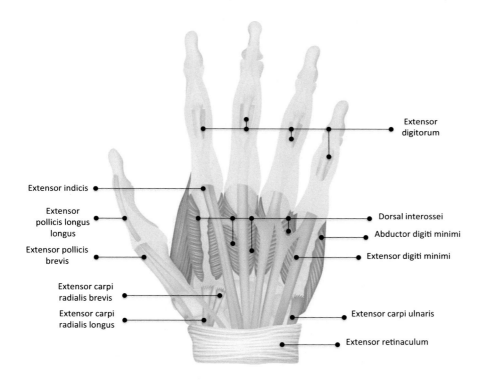

Joint injections

The wrist joint is a complex articulation of the distal radius/ulnar, the carpal bones and the metacarpals. Injection into the dorsal wrist joint (DWJ) usually enables the fluid to dissipate more widely. The 1st carpometacarpal (CMCJ1), metacarpophalangeal (MCPJ), interphalangeal (IPJ) joints of the hand tend to be smaller and less amenable to large volumes; thus more concentrated solutions are beneficial.

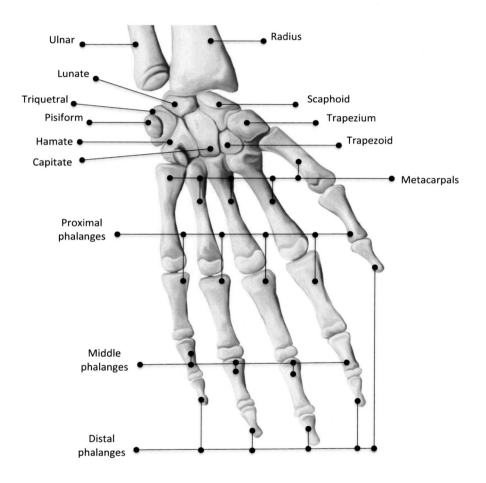

8.3.1 Dorsal wrist joint

Patient position:	In the supine position, the patient can rest their palm flat on the surface of the treatment couch, exposing the dorsum of the hand (Fig. 8.3.1A). In the sitting position, a similar position can be used to access the area while resting the hand on the couch (Fig. 8.3.1B).
Identifying the anatomy:	The dorsal wrist joint (DWJ) can be identified in an LAX view using the base of the 3rd metacarpal as a landmark and identifying the capitate, lunate and distal radius in the same window (Fig. 8.3.1C and D). An SAX view is not usually needed for accessing the DWJ.
Injections performed:	CSI for synovitis, degenerative disease or acute pain. PRP or HA injections for degenerative disease in the wrist.
Recommended transducer:	Hockey stick 8–18 mHz. Linear 6–15 mHz.
Equipment required:	*Equipment preparation:* Set 1 for CSI and HA injections. Set 4 for PRP injections
	Needle: 1- to 1.5-inch 25- or 27-gauge needle.
	Syringes: 3 mL for CSI.
	Medication: 20 mg triamcinolone (0.5 mL) and 1% lidocaine (0.5 mL) for standard CSI.
	Standard/available HA or PRP preparation.
Injection technique:	The injection can be performed in the LAX view using an IP approach with the needle angled at approximately 45–50 degrees (Fig. 8.3.1E and F). The bevel should be facing downwards and the needle can be brought to rest within the joint. A small amount of fluid can be injected to open the space and the needle can be repositioned if needed. Care should be taken to avoid injury to extensor tendons when performing the injection. An SAX view can be used if preferred with an IP approach, but this may be more technically challenging.

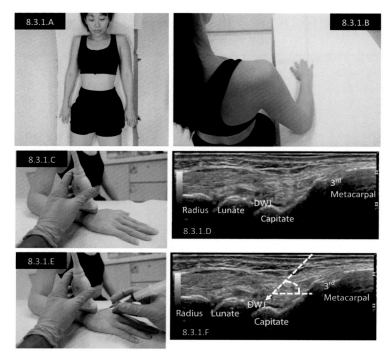

FIGURE 8.3.1 Injections to the DWJ.

In both the lying and seated position, the patient can place their hand prone onto the treatment couch, exposing the dorsal aspect of the wrist. The injection can be undertaken using an IP approach, from distal-to-proximal, with the transducer in a LAX orientation. The needle tip should be seen within mid-carpal space.

8.3.2 1st Carpo-metacarpal joint

Patient position:	In the supine or sitting position, the patient can rest their hand and wrist in a semi-supinated position on the towel or pillow placed on the treatment couch (Fig. 8.3.2A and B). This enables the hand to be taken into a slight ulnar deviation, opening the 1st carpo-metacarpal joint (CMCJ1) and making it more superficial.
Identifying the anatomy:	The 1st CMCJ can be identified in the LAX by tracing the 1st metacarpal proximally and identifying the joint between this and the trapezium (Fig. 8.3.2C and D). Degenerative changes or synovial thickening may be identified here. Placing the wrist into ulnar deviation helps open up the joint.
Injections performed:	CSI for pain, degenerative disease or synovitis. PRP for degenerative joint disease. HA can be attempted but viscosity may limit the injection.
Recommended transducer:	Hockey stick 8–18 mHz.
Equipment required:	*Equipment preparation:* Set 1 for CSI and HA injections. Set 4 for PRP injections. *Needle:* 1- to 1.5-inch 25- or 27-gauge needle. *Syringes:* 3 mL for CSI. *Medication:* 20 mg triamcinolone (0.5 mL) and 1% lidocaine (0.5 mL) for standard CSI. Standard/available HA or PRP preparation.
Injection technique:	The injection is best performed in the LAX orientation using an IP approach from proximal to distal. The needle should be angled at approximately 30 degrees (Fig. 8.3.2E and F), with the bevel facing down and the tip aimed to rest within the joint itself. Care should be taken to avoid injury to tendons when performing the injection. Once within the joint, there may be a build-up of pressure before this eases and flow is noted. An OOP approach can also be used to guide the needle, with the transducer placed across the joint and the needle inserted next to it and almost perpendicular the skin (Fig. 8.3.2G and H).

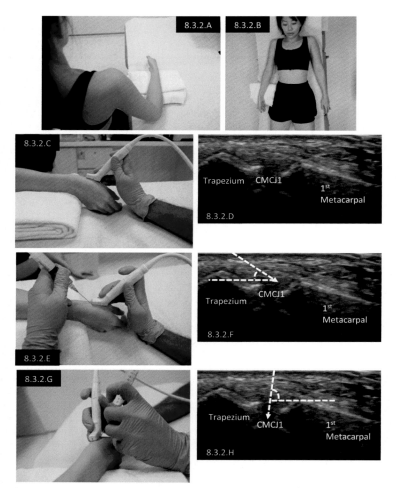

FIGURE 8.3.2 Injections to the CMCJ1.

In both the lying and seated position, the patient can place their forearm into a semi-supinated position with the wrist on a rolled towel, enabling the hand to be taken into ulnar deviation. The injection can be undertaken using an IP approach, from proximal-to-distal, or using an OOP technique; for both the transducer is in a LAX orientation. The needle tip should be seen within the CMCJ1.

8.3.3 **Metacarpo-phalangeal joint**

Patient position:	In the supine or sitting position, the patient can rest their hand and wrist in a prone position on the treatment couch (Fig. 8.3.3A and B). This enables the metacarpo-phalangeal (MCPJ) and inter-phalangeal joints (IPJ) to be viewed from the dorsal aspect.
Identifying the anatomy:	Tracing the dorsal surface of the metacarpal bones from a proximal to distal orientation, the various MCPJs and IPJs can be identified in a LAX view (Fig. 8.3.3C and D). Degenerative changes might be seen as cortical irregularities or synovial thickening may signify synovitis.
Injections performed:	CSI for pain, degenerative disease or synovitis. PRP for degenerative joint disease. HA can be attempted but viscosity may limit the injection.
Recommended transducer:	Hockey stick 8–18 mHz.
Equipment required:	*Equipment preparation:* Set 1 for CSI and HA injections. Set 4 for PRP injections. *Needle:* 1-inch 25- or 27-gauge needle. *Syringes:* 3 mL for CSI. *Medication:* 20 mg triamcinolone (0.5 mL) and 1% lidocaine (0.5 mL) for standard CSI. Standard/available HA or PRP preparation.
Injection technique:	The injection can be performed with the transducer in a LAX orientation using a proximal-to-distal IP approach. The needle is angled at approximately 20 degrees (Fig. 8.3.3E and F), with the bevel facing down and the tip aimed to rest within the joint itself. Alternatively, an OOP approach can be taken with the transducer over the joint and the needle inserted just next to its midline and perpendicular to the skin (Fig. 8.3.3G and H). Care should be taken to avoid injury to tendons and neurovascular structures when performing the injection.

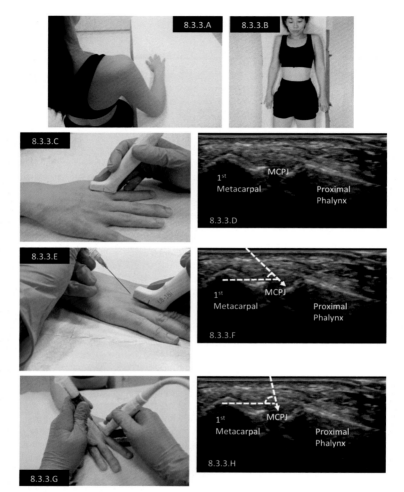

FIGURE 8.3.3 Injections to the MCPJ.

In both the lying and seated position, the patient can place their hand prone on the treatment couch. The injection can be undertaken using an IP or OOP approach, with the transducer in a LAX orientation. The needle tip should be seen within the MCPJ.

Tendon injections

Common tendon pathologies around the hand and wrist include De Quervain's teno-synovitis (DQT), extensor carpi ulnaris (ECU), tendinopathies, intersection syn-dromes (IS) and trigger fingers (TF).

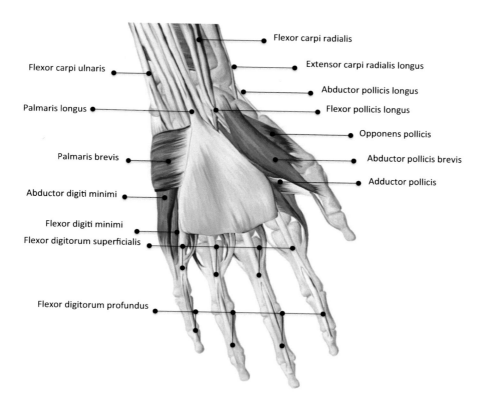

8.3.4 De Quervain's tenosynovitis

Patient position:	In the supine or seated position, the patient can rest their hand on the surface of the treatment couch or a towel, in a semi-supinated position, exposing the radial border of the wrist (Fig. 8.3.4A and B). This makes the abductor pollicis longus (APL) and extensor pollicis brevis (EPB) tendons easily accessible.
Identifying the anatomy:	Identifying Lister's tubercle and tracing to the radial border of the wrist help locate the first extensor compartment. Once in position the De Quervain's tenosynovitis (DQT) tendons (APL and EPB) can be traced distally in an LAX (Fig. 8.3.4C and D) or SAX view and oedema and thickening of the teno-synovium maybe noted (Fig. 8.3.4E and F).
Injections performed:	CSI for tenosynovitis. PRP for degenerative tendon disease with intra-substance tearing.
Recommended transducer:	Hockey stick 8–18 mHz. Linear 6–15 mHz.
Equipment required:	*Equipment preparation:* Set 1 for CSI. Set 4 for PRP injections. *Needle:* 1- to 1.5-inch 25- or 27-gauge needle. *Syringes:* 3 mL for CSI. *Medication:* 20 mg triamcinolone (1 mL) and 1% lidocaine (0.5 mL) for standard CSI. Standard/available PRP preparations.
Injection technique:	The injection can be performed in the LAX or SAX view using an IP approach with the needle angled at approximately 20 degrees in the former (Fig. 8.3.4G and H) and as horizontal as possible in the latter (Fig. 8.3.4I and J). With the bevel down, the needle can be brought to rest against the tendon within the sheath and once there is separation of the two planes, the needle can be slightly withdrawn and the fluid will be seen flowing in the space created. Care should be taken to avoid injecting directly into the tendons and injury to nerves or vessels. An OOP view can confirm the positioning of the needle in the space created after injecting the solution. For PRP injections, the needle is passed into the tendon itself and the solution is distributed using a fenestration technique.

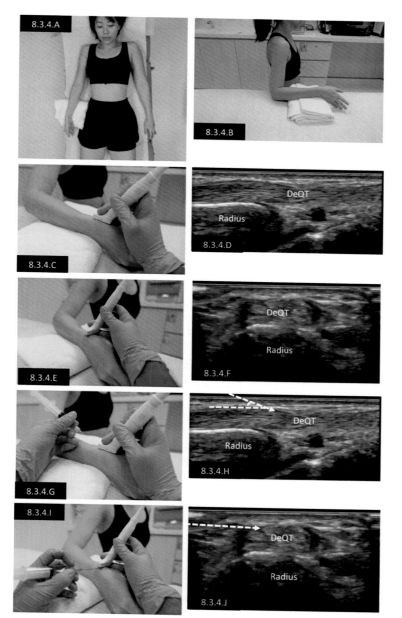

FIGURE 8.3.4 Injections to the DQT.

In both the lying and seated position, the patient can place their forearm into a semi-supinated position with the wrist on a rolled towel, enabling the hand to be taken into ulnar deviation. The injection can be undertaken using an IP approach, from proximal-to-distal or lateral-to-medial, with the transducer in a LAX or SAX orientation. The needle tip should be seen between the tendons and its sheath.

8.3.5 Extensor carpi ulnaris

Patient position:	In the supine or sitting position, the patient can rest their hand on the surface of the treatment couch, in an overpronated position, exposing the ulnar border of the wrist and bringing the extensor carpi ulnaris (ECU) tendon into an easily accessible position (Fig. 8.3.5A and B).
Identifying the anatomy:	Identifying Lister's tubercle and then tracing to the ulnar border helps locate the sixth extensor compartment (Fig. 8.3.5C and D). Once in position, the ECU tendon can be traced distally in the ulnar groove with the retinaculum overlying. The tendon can also be assessed in the LAX view for thickening and vascularity (Fig. 8.3.5E and F).
Injections performed:	CSI for pain from tendinopathy. PRP for degenerative tendon disease with intrasubstance tearing.
Recommended transducer:	Hockey stick 8–18 mHz. Linear 6–15 mHz.
Equipment required:	*Equipment preparation:* Set 1 for CSI. Set 4 for PRP injections. *Needle:* 1- to 1.5-inch 25- or 27-gauge needle. *Syringes:* 3 mL for CSI. *Medication:* 20 mg triamcinolone (1 mL) and 1% lidocaine (0.5 mL) for standard CSI. Standard/available PRP preparations.
Injection technique:	The injection can be performed in a SAX (Fig. 8.3.5G and H) or LAX (Fig. 8.3.5I and J) view using an IP approach with the needle as horizontal as possible for the former and angled at 20 degrees for the latter. With the bevel down, the needle can be brought to rest against the tendon within the sheath and once there is separation of the tissue planes, it can be re-adjusted if needed. The fluid should be seen flowing around the tendon. Care should be taken to avoid injecting directly into the tendons. For PRP injections, the needle is passed into the tendon itself and the solution is distributed using a fenestration technique.

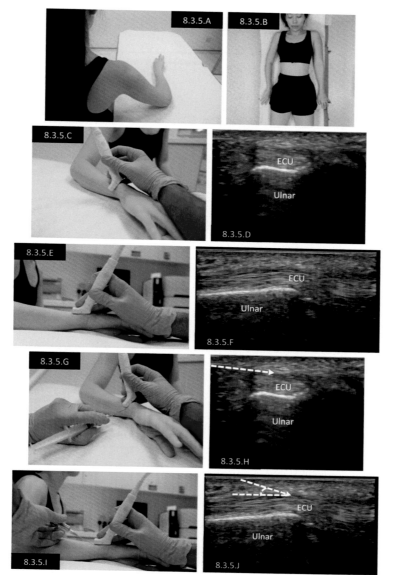

FIGURE 8.3.5 Injections to the ECU.

In both the lying and seated position, the patient can place their forearm into an overpronated position. The injection can be undertaken using an IP approach, from proximal-to-distal or lateral-to-medial, with the transducer in a LAX or SAX orientation. The needle tip should be seen between the tendon and its sheath.

8.3.6 Intersection syndrome

Patient position:	In the supine or sitting position, the patient can rest their palm flat on the surface of the treatment couch or resting on a pillow. The hand can be kept in a pronated position, exposing the dorsal aspect of the wrist and forearm (Fig. 8.3.6A and B).
Identifying the anatomy:	Two intersections can be identified in an SAX view. The proximal involves the first compartment (APL/EPB) crossing over the second (ECRB/ERCL) (Fig. 8.3.6C and D) while the distal one involves the third compartment (EPL) crossing over the second (Fig. 8.3.6E and F).
Injections performed:	CSI for intersection syndrome.
Recommended transducer:	Hockey stick 8—18 mHz. Linear 6—15 mHz.
Equipment required:	*Equipment preparation:* Set 1 for CSI. *Needle:* 1- to 1.5-inch 25- or 27-gauge needle. *Syringes:* 3 mL for CSI. *Medication:* 20 mg triamcinolone (1 mL) and 1% lidocaine (0.5 mL) for standard CSI injections.
Injection technique:	The injection is easiest performed in the SAX view using an IP approach with the needle angled at 20—30 degrees (*proximal:* Fig. 8.3.6G and H; *distal:* Fig. 8.3.6I and J). With the bevel facing down, the needle is positioned in the plane between the two compartments before a small amount of fluid is injected into the space to cause a separation of the tissue planes. The needle can then be repositioned if required, before the solution is fully injected. Care should be taken to avoid injecting directly into the tendons and nerves.

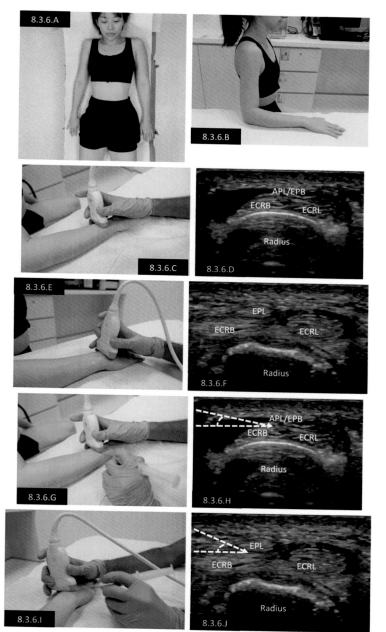

FIGURE 8.3.6 Injections for intersection syndrome.

In both the lying and seated position, the patient can place their forearm into a pronated position with the hand resting on the treatment couch. The injection can be undertaken using an IP approach, from lateral-to-medial, with the transducer in a SAX orientation. The needle tip should be seen between the tendons.

8.3.7 Trigger finger

Patient position:	In the supine or seated position, the patient can rest their hand flat on the surface of the treatment couch, in a supinated position, exposing the palmar aspect (Fig. 8.3.7A and B).
Identifying the anatomy:	The finger in question can be examined in an SAX or LAX view. In the SAX view, a hypoechogenic area around the tendon can be identified surrounding the tendon, suggestive of a thickened pulley (Fig. 8.3.7C). In the latter, a prominent pulley maybe identified and if the finger is flexed, it can be seen to catch after a certain degree of movement (Fig. 8.3.7D).
Injections performed:	CSI for pain and triggering.
Recommended transducer:	Hockey stick 8–18 mHz. Linear 6–15 mHz.
Equipment required:	*Equipment preparation:* Set 1 for CSI. *Needle:* 1- to 1.5-inch 25- or 27-gauge needle. *Syringes:* 3 mL for CSI. *Medication:* 20 mg triamcinolone (1 mL) and 1% lidocaine (0.5 mL) for standard CSI.
Injection technique:	The trigger finger injection is easiest performed in the SAX, using an IP approach with the needle angled at 20 degrees (Fig. 8.3.7E and F). With the bevel down, the needle it is positioned in the plane between the pulley and the tendon and once this space opens, it can be repositioned if needed. Fluid should be seen flowing around the tendon and the patient might report a tightness sensation in the digit. An LAX orientation can also be used in a proximal-to-distal orientation (Fig. 8.3.7I and J). Care should be taken to avoid injecting directly into the tendons and any neurovascular structures nearby.

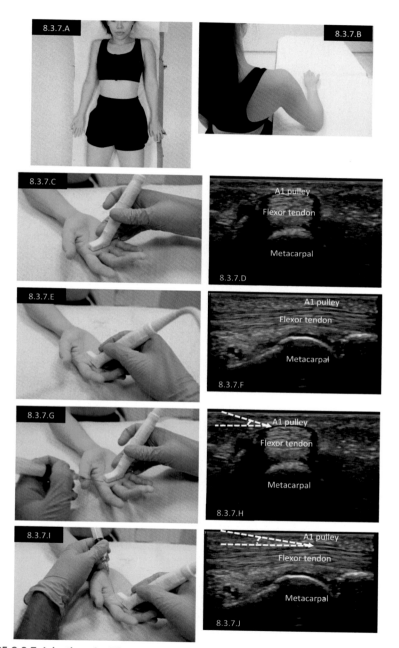

FIGURE 8.3.7 Injections for TF.

In both the lying and seated position, the patient can place their forearm into a supinated position, with hand resting on the treatment couch. The injection can be undertaken using an IP approach, with the transducer in a LAX or SAX orientation. The needle tip should be seen between the tendon and the A1 pulley.

Nerve injections

The median nerve (MN) is commonly affected as it passes beneath the flexor retinaculum into the carpal tunnel on the volar aspect of the wrist. Being superficial, it can be assessed on its path in the forearm and it is a common nerve injection undertaken in the wrist.

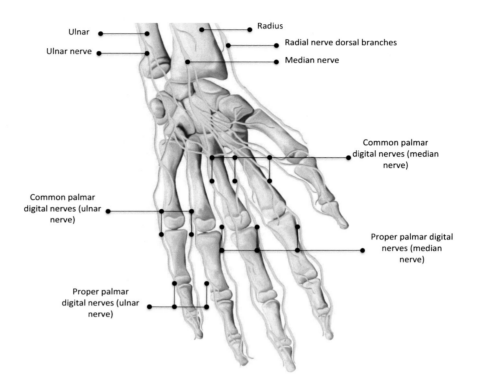

Ulnar

Ulnar nerve

Radius

Radial nerve dorsal branches

Median nerve

Common palmar digital nerves (median nerve)

Common palmar digital nerves (ulnar nerve)

Proper palmar digital nerves (median nerve)

Proper palmar digital nerves (ulnar nerve)

8.3.8 Median nerve

Patient position:	In the supine or seated position, the patient can rest their wrist flat on the surface of the treatment couch, with the hand in a supinated position, exposing the palmar aspect of the wrist and forearm (Fig. 8.3.8A and B). A pillow or rolled towel can be used for patient comfort or stability.
Identifying the anatomy:	With a bunch of grapes appearance, the median nerve (MN) can be seen travelling under the flexor retinaculum into the carpal tunnel in the SAX (Fig. 8.3.8C and D), before it divides into smaller branches. In this view, the cross-sectional circumference can be measured and if greater than 10 mm squared, this is indicative of carpal tunnel syndrome (CTS). Thickening of the flexor retinaculum can also be identified.
Injections performed:	CSI for pain and neuropathy.
Recommended transducer:	Hockey stick 8—18 mHz. Linear 6—15 mHz.
Equipment required:	*Equipment preparation:* Set 1 for CSI. *Needle:* 1- to 1.5-inch 25- or 27-gauge needle. *Syringes:* 3 mL for CSI. *Medication:* 20 mg triamcinolone (1 mL) and 1% lidocaine (0.5 mL) for standard CSI.
Injection technique:	The injection is easiest performed in the SAX using an IP approach with the needle angled at 20—30 degrees (Fig. 8.3.8E and F). With the bevel down, the needle is positioned between the MN and flexor retinaculum before fluid is injected to cause a separation of the tissue planes. Once margins are more clearly defined, the needle can be repositioned by angling more steeply to beneath the MN. Here, further CSI can be injected.

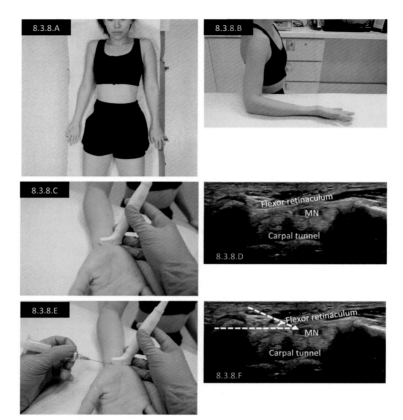

FIGURE 8.3.8 Injections for CTS.

In both the lying and seated position, the patient can place their forearm into a supinated position, with hand resting on the treatment couch. The injection can be undertaken using an IP approach, with the transducer in a SAX orientation. The needle tip should be seen between the MN and flexor retinaculum.

Ligament injections

The scapholunate ligament (SLL) in the wrist can be commonly injured following falls onto the outstretched hand. In situations where there is a pain or partial tears, a guided injection can help with healing and functional recovery.

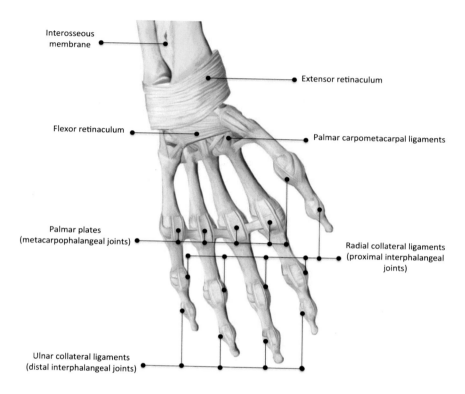

Interosseous membrane

Extensor retinaculum

Flexor retinaculum

Palmar carpometacarpal ligaments

Palmar plates (metacarpophalangeal joints)

Radial collateral ligaments (proximal interphalangeal joints)

Ulnar collateral ligaments (distal interphalangeal joints)

8.3.9 Scapholunate ligament

Patient position:	In the supine or seated position, the patient can rest their wrist flat on the surface of the treatment couch, exposing the dorsal aspect of the hand (Fig. 8.3.9A and B). If needed, a pillow or towel can be used to support the wrist.
Identifying the anatomy:	The scapholunate ligament (SLL) can be identified in an SAX view by finding Lister's tubercle and moving distally. The ligament should be seen running between the scaphoid and lunate bones immediately distal to Lister's tubercle (Fig. 8.3.9C and D). Alternatively, the lunate can be identified in the LAX view and then the transducer can be rotated 90° so that the ligament is brought into view.
Injections performed:	CSI for pain symptoms. Prolo for partial ligament tears.
Recommended transducer:	Hockey stick 8–18 mHz. Linear 6–15 mHz.
Equipment required:	*Equipment preparation:* Set 1 for CSI. Set 5 for Prolo. *Needle:* 1- to 1.5-inch 25- or 27-gauge needle. *Syringes:* 3 mL for CSI. *Medication:* 20 mg triamcinolone (0.5 mL) and 1% lidocaine (0.5 mL) for standard CSI. For Prolo, 2 mL of a 50:50 mixture of 50% dextrose and 1% lidocaine can be used.
Injection technique:	The injection is easiest performed in the SAX view using an IP approach with the needle angled at approximately 30 degrees (Fig. 8.3.9E and F). With the bevel down, the needle can be brought to rest above the ligament for CSI and the solution should be injected above the ligament. Once the tissue planes separate, the needle can be repositioned if needed, to avoid injecting into the ligament itself. For Prolo injections, the needle can be introduced into the ligament and a fenestration technique can be used when injecting the solution to distribute it. Care should be taken to avoid injury to tendons when performing the injection.

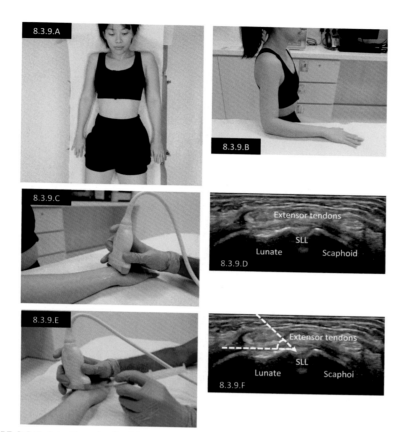

FIGURE 8.3.9 Injections to the SLL.

In both the lying and seated position, the patient can place their forearm into a pronated position, with hand resting on the treatment couch. The injection can be undertaken using an IP approach, from the medial or lateral aspect, with the transducer in a SAX orientation. The needle tip should be seen within the ligament.

Notes
(Please use this area to reflect on your procedure and how you can build on these experiences).

Spine

While detailed assessments require imaging such as MRI, with many structures around the spine relatively superficial, ultrasound can be used to provide peri-radicular nerve root (NR) and facet joint (FJ) injections in the lumbar and cervical spine. In addition, the sacro-iliac joints (SIJ) can be visualised for treatments.

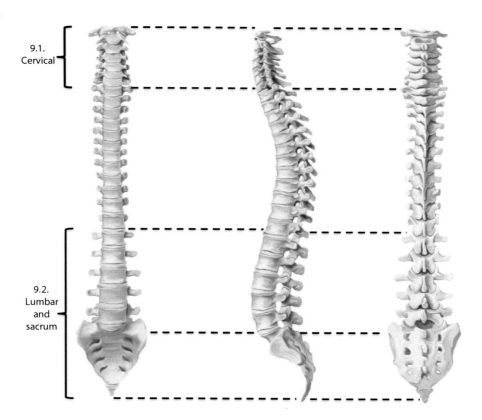

9.1. Cervical

9.2. Lumbar and sacrum

9.1 **Cervical spine**

Traditionally, cervical facet joints (CFJ) and nerve roots (CNR) have been injected under fluoroscopic or CT guidance. The advent of real-time ultrasound has enabled this to be done with efficiency, safety and without the need for exposing the patient to undue radiation. The other advantage of ultrasound is that vascular structures can be visualized using doppler flow.

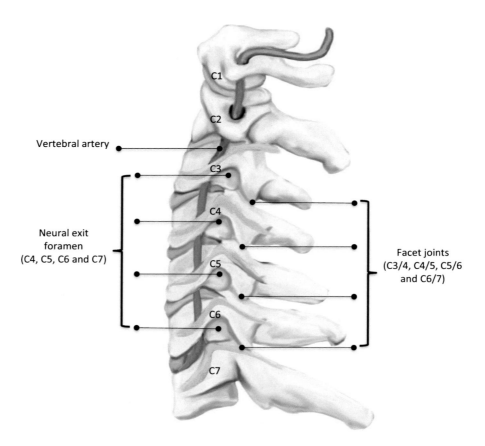

Joints

CFJ can be identified at the posterior aspect of the neck and injections can be undertaken using ultrasound guidance. In situations of diagnostic uncertainty, LA alone can be used, but for longer therapeutic duration steroid can be added.

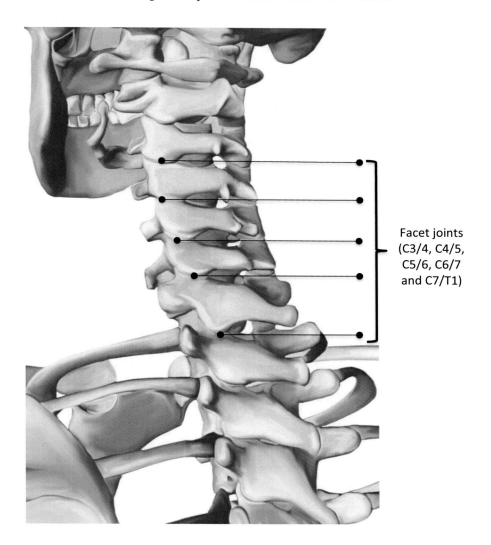

Facet joints (C3/4, C4/5, C5/6, C6/7 and C7/T1)

9.1.1 Cervical Facet Joints

Patient position:	For the cervical facet joints (CFJ), the patient can be positioned in a side-lying position with the neck slightly side flexed to the contralateral side (Fig. 9.1.1A). This opens up the side requiring treatment, particularly in patients with shorter necks. Alternatively, the patient can be in a sitting position with the neck in a neutral or slightly extended position (Fig. 9.1.1B).
Identifying the anatomy:	Placed on the lateral aspect of the neck with the transducer in a SAX view, the C7 vertebra can be identified with its single posterior process. The C7/T1 facet joint can be seen by moving the transducer posteriorly and as the lamina comes into view the joint can be identified (Fig. 9.1.1C–E). The C6 level is identified by moving transducer proximally. It has two lateral processes and the C6 CFJ can be found by moving the transducer posteriorly. Higher levels can be identified in a similar manner.
Injections performed:	CSI for CFJ pain.
Recommended transducer:	Linear 8–12 mHz.
Equipment suggested:	*Equipment preparation:* Set 6 for facet, nerve root and sacro-iliac joint injections. *Needle:* 1.5- to 2.0-inch 23- or 25-gauge needle. *Syringes:* 3 mL for CSI. *Medication:* 2 mg dexamethasone (0.5 mL) and 1% lidocaine (1.5 mL) for standard CSI.
Injection technique:	Viewing the CFJ in the SAX, the needle can be introduced from a posterior position using an IP technique. The angle depends on body habitus but will be generally around 30–45 degrees (Fig. 9.1.1F–H). The needle tip can be brought to rest over the joint itself, with the bevel facing down, and once no blood flow is confirmed, the solution can be injected. The solution might not be seen flowing into the joint itself, but if delivered as close to is as possible, the active components will still have a therapeutic effect.

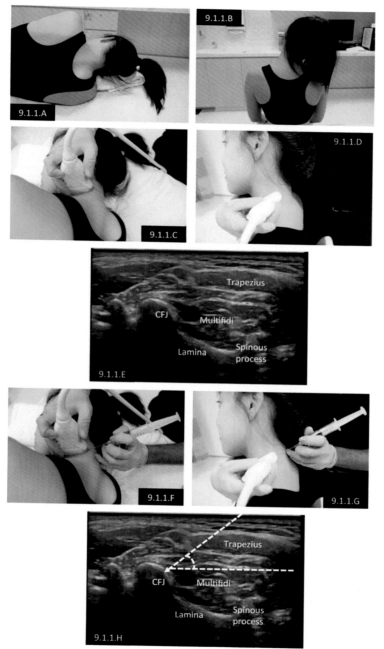

FIGURE 9.1.1 Injections to the CFJ.

In the side-lying position the neck is side flexed to the contralateral side and the injection can be undertaken with the transducer in a SAX orientation over the neck. The CFJ is seen in a SAX view. The needle is introduced IP from a posterior approach. Care must be taken not to damage nearby neurovascular structures.

Nerves

The cervical nerve roots (CNR) can be visualised and injected under ultrasound guidance with a high degree of reliability and safety.

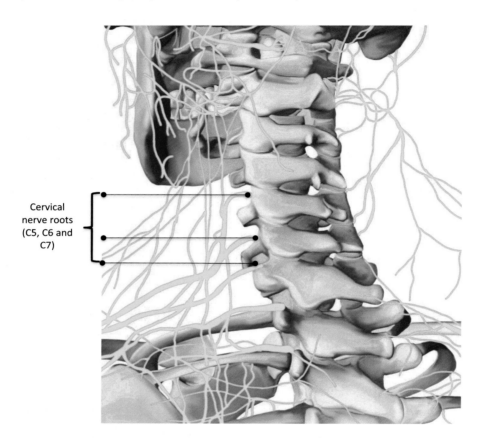

Cervical nerve roots (C5, C6 and C7)

9.1.2 **Cervical Nerve Roots (Peri-radicular)**

Patient position:	For the cervical nerve roots (CNR), the patient can be placed in a side-lying position with the neck side flexed to the contralateral side (Fig. 9.1.2A). This opens the side requiring treatment and is particularly useful in patients with shorter necks. Alternatively, the patient can be in a sitting position with the neck neutral or slightly extended (Fig. 9.1.2B).
Identifying the anatomy:	Placed on the lateral aspect of the neck and the transducer in a SAX view (Fig. 9.1.2C and D), the C7 vertebra can be identified by the single, lateral process and the CNR can be seen exiting as the transducer is moved anteriorly (Fig. 9.1.2E). Similarly, the C6 CNR above this can be identified by locating the C6 level, which has an anterior and posterior process, and the nerve is situated between both processes (Fig. 9.1.2F). Levels above C6 will have an anterior and posterior process thereafter and the nerves will be seen between these. At levels above C7, it is important to apply the power doppler to identify where the vertebral artery is situated to avoid injury during the procedure.
Injections performed:	CSI for radicular pain (lidocaine is omitted).
Recommended transducer:	Linear 8–12 mHz.
Equipment suggested:	*Equipment preparation:* Set 6 for facet, nerve root and sacro-iliac joint injections. *Needle:* 1.5- to 2.0-inch 23- or 25-gauge needle. *Syringes:* 3 mL for CSI. *Medication:* 2 mg dexamethasone (0.5 mL) only.
Injection technique:	Viewing the CNR in the SAX, the needle can be introduced from a posterior position using an IP technique (Fig. 9.1.2G and H). The angle depends on body habitus but will be generally around 50–60 degrees (Fig. 9.1.2I and J). The needle tip can be brought to rest above and slightly posterior to the nerve with the bevel facing down. It is useful to have the power Doppler function active to ensure the vertebral blood vessel is not nearby prior to undertaking the procedure. If there is no flow, then the solution can be safely injected.

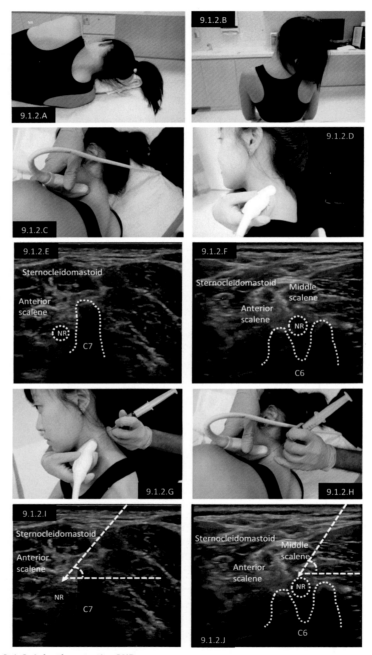

FIGURE 9.1.2 Injections to the CNR.

In the side-lying position the neck is side flexed to the contralateral side and the injection can be undertaken with the transducer in a SAX orientation over the neck. The needle is introduced IP from a posterior approach to near the CNR and care must be taken not to damage the nearby vascular structures.

9.2 **Lumbar Spine and Sacrum**

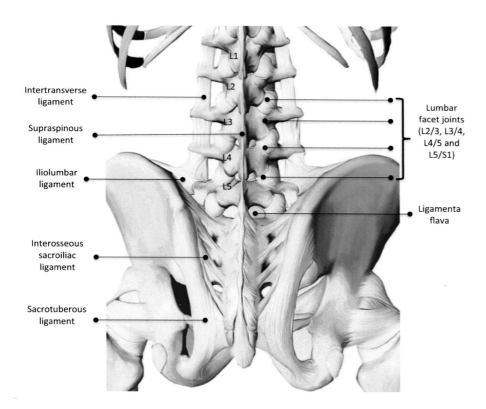

Intertransverse ligament

Supraspinous ligament

Iliolumbar ligament

Interosseous sacroiliac ligament

Sacrotuberous ligament

Lumbar facet joints (L2/3, L3/4, L4/5 and L5/S1)

Ligamenta flava

L1
L2
L3
L4
L5

Joint Injections

Lumbar facet joints (LFJ) and the SIJ can be injected using ultrasound guidance. The real-time imaging provided by ultrasound enable rapid adjustments and even with older or larger patients, images can be optimised to enable visualisation.

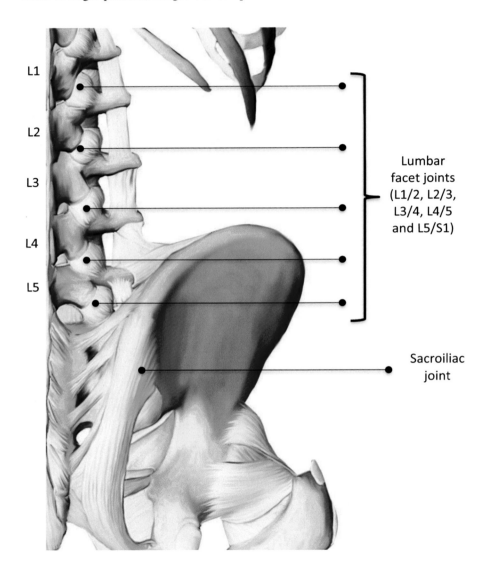

L1

L2

L3

L4

L5

Lumbar
facet joints
(L1/2, L2/3,
L3/4, L4/5
and L5/S1)

Sacroiliac
joint

9.2.1 **Lumbar Facet Joints**

Patient position:	For lumbar facet joints (LFJ), the patient should be positioned in a prone position with the legs and upper body taken into a slightly flexed position by dropping the ends of the treatment couch (Fig. 9.2.1A and B). The lower back must be exposed from mid-thoracic spine to natal cleft.
Identifying the anatomy:	Placed in an SAX view over the lower back, the Sacrum can be identified with small spinous processes, while the lumbar have relatively taller processes. Once the lumbar area and spinous process is identified, the transducer should be moved laterally and kept in an SAX view. The facet joint is identified as a break in the cortical continuity between the lamina of the superior vertebra and the pedicle of the inferior one (Fig. 9.2.1C and D).
Injections performed:	CSI for LFJ pain.
Recommended transducer:	Curvilinear 2–5 mHz.
Equipment suggested:	*Equipment preparation:* Set 6 for facet, nerve root and sacro-iliac joint injections. *Needle:* 2.0- to 2.5-inch 21- or 23-gauge needle. *Syringes:* 3 mL for CSI. *Medication:* 2 mg dexamethasone (0.5 mL) and 1% lidocaine (1.5 mL) for standard CSI.
Injection technique:	Viewing the LFJ in the SAX, the needle can be introduced from a lateral position using an IP technique. The angle depends on body habitus but will be generally around 60 degrees (Fig. 9.2.1E and F). The needle tip can be brought to rest over the joint itself, with the bevel facing down, and once no blood flow is confirmed, the solution can be injected.

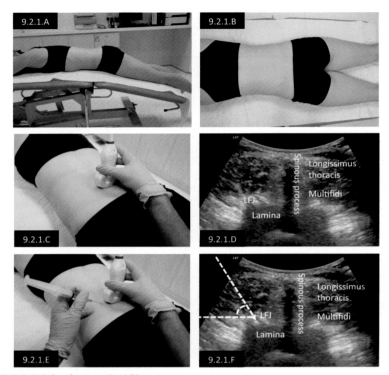

FIGURE 9.2.1 Injections to the LFJ.

In the prone position the hips are flexed to make the lumbosacral spine more superficial. The injection can be undertaken with the transducer in a SAX orientation over the LFJ and the needle is introduced IP from a lateral approach.

9.2.2 **Sacro-Iliac Joints**

Patient position:	For the sacro-iliac joints (SIJ), the patient should be positioned in a prone position with the legs and upper body taken into a slightly flexed position by dropping the ends of the treatment couch (Fig. 9.2.2A and B). The lower back must be exposed from mid-thoracic spine to natal cleft.
Identifying the anatomy:	Placed in an SAX view over the lower back, the sacrum can be identified with small spinous processes, while the lumbar have relatively taller processes. Keeping the transducer over the Sacrum, moving the transducer laterally and in an SAX view, the Ilium can be seen and the junction between them is where the SIJ is located (Fig. 9.2.2C and D).
Injections performed:	CSI for pain, degenerative disease or synovitis. PRP injections for pain and stability concerns.
Recommended transducer:	Curvilinear 2—5 mHz.
Equipment suggested:	*Equipment preparation:* Set 6 for facet, nerve root injections and SIJ. Set 4 for PRP injections. *Needle:* 2.0- to 2.5-inch 21- or 23-gauge needle. *Syringes:* 3 mL for CSI. *Medication:* 2 mg dexamethasone (0.5 mL) and 1% lidocaine (1.5 mL) for standard CSI. Standard/available PRP preparation.
Injection technique:	Viewing the SIJ in the SAX, the needle can be introduced from a medial position over the spine of the Sacrum, using an IP technique (Fig. 9.2.2E and F). The angle of approach depends on body habitus but will be generally around 45—50 degrees. The needle tip can be brought to rest in the joint itself, with the bevel facing down, and once no blood flow is confirmed, the solution can be injected.

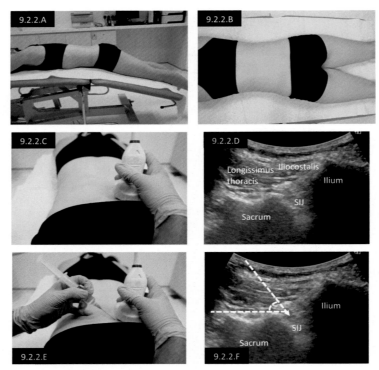

FIGURE 9.2.2 Injections to the SIJ.

In the prone position the hips are flexed to make the lumbosacral spine more superficial. The injection can be undertaken with the transducer in a SAX orientation over the SIJ and the needle is introduced IP from a medial approach to beneath the ligament and into the joint.

Nerves

Lumbar nerves (LN) and sacral nerves (SN) can be injected in a peri-radicular manner using ultrasound guidance. Once placed around the nerve, the steroid will diffuse to the area of inflammation.

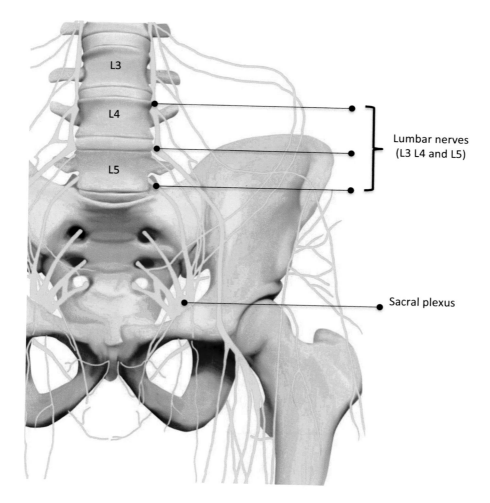

9.2.3 **Lumbar Nerves**

Patient position:	For peri-radicular lumbar nerve (LN) injections, the patient can be positioned in a prone position with the legs and upper body taken into a slightly flexed position by lowering the ends of the treatment couch (Fig. 9.2.3A and B). The lower back must be exposed from mid-thoracic spine to natal cleft.
Identifying the anatomy:	Placed in an SAX view over the lower back, the Sacrum can be identified with small spinous processes, while the lumbar have relatively taller processes. Once identified, the transducer can be rotated 90 degrees and placed in an LAX position with the spinous processes aligned if possible. The transducer is then moved laterally to visualize the transverse processes (TP) in an SAX view (Fig. 9.2.3C and D). The exiting LN are situated beneath the intertransverse ligament (ITL) that courses between the TP.
Injections performed:	CSI for radicular pain (lidocaine is omitted).
Recommended transducer:	Curvilinear 2–5 mHz.
Equipment suggested:	*Equipment preparation:* Set 6 for facet, nerve root injections and SIJ. *Needle:* 2.0- to 2.5-inch 21- or 23-gauge needle. *Syringes:* 3 mL for CSI. *Medication:* 2 mg dexamethasone (0.5 mL) only.
Injection technique:	Viewing the TP in the SAX, the needle can be introduced from a cranial to caudal orientation using an IP technique. The angle of approach depends on body habitus but will be generally around 60–70 degrees. With the bevel facing down, the needle tip must be guided through the ITL to just below the plane of the TP (Fig. 9.2.3E and F). Once no blood flow is confirmed, the solution can be injected. The challenge with this procedure is maintaining the needle in the IP view, particularly with the depth of the injection.

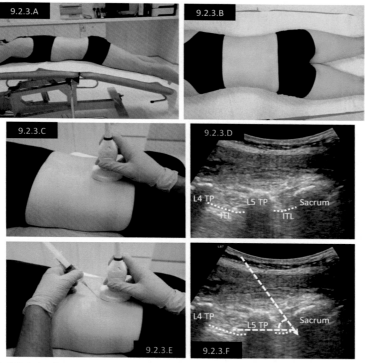

FIGURE 9.2.3 Injections to the LNR.

In the prone position the hips are flexed to make the lumbosacral spine more superficial. The injection can be undertaken with the transducer in a LAX orientation over lumbar spine and the TP are in a SAX view. The needle is introduced IP from a cranial approach to beneath the transverse ligament and around the exiting LNR.

9.2.4 Sacral Nerves

Patient position:	For the sacral nerves (SN), the patient should be positioned in a prone position with the legs and upper body taken into a slightly flexed position by dropping the ends of the treatment couch (Fig. 9.2.4A and B). The lower back must be exposed from mid-thoracic spine to natal cleft.
Identifying the anatomy:	Placed in an SAX view over the lower back, the sacrum can be identified with small spinous processes, while the lumbar have relatively taller processes. Maintaining the transducer over the sacrum and moving it distally and laterally, the Sacral Foramina can be visualised as a dip in the sacral cortex and in increase in reflective shadowing (Fig. 9.2.4C and D).
Injections performed:	CSI for radicular pain (lidocaine is omitted).
Recommended transducer:	Curvilinear 2–5 mHz.
Equipment suggested:	*Equipment preparation:* Set 6 for facet, nerve root injections and SIJ. *Needle:* 2.0- to 2.5-inch 21- or 23-gauge needle. *Syringes:* 3 mL for H&L injection. *Medication:* 2 mg dexamethasone (0.5 mL) only.
Injection technique:	Viewing the SIJ in the SAX, the needle can be introduced from a medial position over the spine of the sacrum, using an IP technique. The angle depends on body habitus but will be generally around 50–60 degrees (Fig. 9.2.4E and F). With the bevel facing down, the needle tip can be brought to rest above the foramen itself and once no blood flow is confirmed, the solution can be injected and should be seen flowing into the foramen. Alternatively, an OOP approach can be used with the transducer over the joint and the needle inserted adjacent to the long border of the transducer and perpendicular to the skin surface (Fig. 9.2.4G and H). Once the needle tip is seen within the foramen and no blood flow is confirmed on aspiration, the injection can be performed.

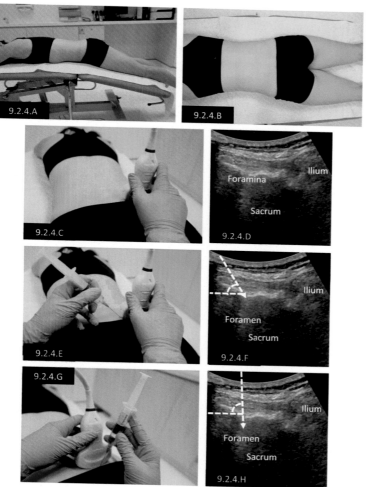

FIGURE 9.2.4 Injections to the SNR.

In the prone position the hips are flexed to make the lumbosacral spine more superficial. The injection can be undertaken with the transducer in a SAX orientation over sacrum spine and the sacral foramen are in a SAX view. The needle is introduced OOP into the foramen and around the exiting SNR.

Notes
(Please use this area to reflect on your procedure and how you can build on these experiences).

Lower limb

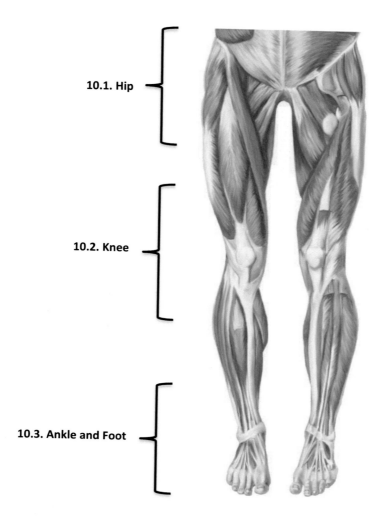

10.1. Hip

10.2. Knee

10.3. Ankle and Foot

Ultrasound Guided Musculoskeletal Procedures in Sports Medicine. https://doi.org/10.1016/B978-0-323-91014-9.00012-0
Copyright © 2021 Elsevier Inc. All rights reserved.

10.1 Hip

The hip represents a challenging area to scan and inject effectively. The key elements in this region are the correct identification of anatomy, suitable positioning of the patient, bring structures as superficial as possible, and adapting techniques to suit the individual clinician. In this way, multiple areas can be effectively treated.

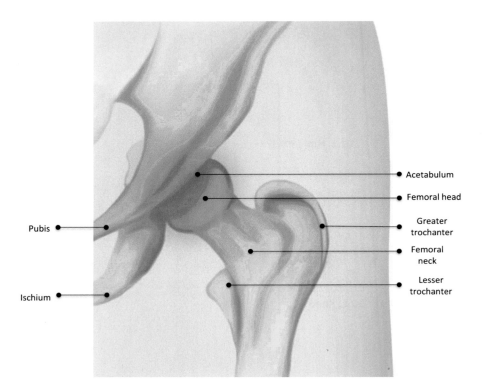

Pubis

Ischium

Acetabulum

Femoral head

Greater trochanter

Femoral neck

Lesser trochanter

Joint injections

Two common joints that might require an injection this area include the hip joint (HJ) itself and the pubic symphysis (PS). Being deeper, the former tends to be more challenging while the latter, although more superficial can be associated with more pain during the procedure.

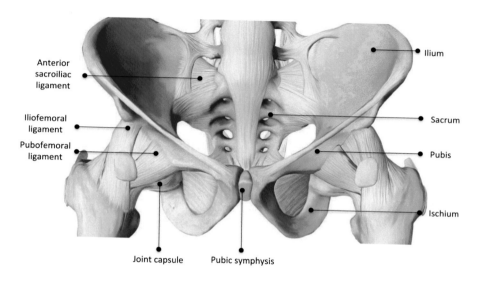

Anterior sacroiliac ligament

Iliofemoral ligament

Pubofemoral ligament

Joint capsule

Pubic symphysis

Ilium

Sacrum

Pubis

Ischium

10.1.1 Anterior Hip joint

Patient position:	For the anterior hip joint (AHJ), the patient should be positioned in a supine orientation and, if possible, the hip can be opened further, by lowering the end of the bed (Fig. 10.1.1A). This is particularly useful if there is extensive degenerative changes in the joint.
Identifying the anatomy:	In this position, the anterior joint is most readily visualised by placing the transducer in a LAX orientation, slightly oblique to the position of the leg. In the field of view, the femoral neck, femoral head and acetabulum should be visualised (Fig. 10.1.1B and C). The labrum may also be identified and in degenerative hips an effusion, bony irregularity or labral cysts might be seen.
Injections performed:	CSI for pain symptoms or synovitis. PRP or HA injections for degenerative disease.
Recommended transducer:	Curvilinear 3—5 mHz.
Equipment suggested:	*Equipment preparation:* Set 1 for CSI or HA injections. Set 4 for PRP. *Needle:* 2.5-inch 21- or 23-gauge needle. *Syringes:* 3 mL for CSI. *Medication:* 40 mg triamcinolone (1 mL) and 1% lidocaine (2 mL). Standard/available HA or PRP preparation.
Injection technique:	Using the femoral neck as the target and the transducer in a LAX orientation, the needle can be inserted using an IP approach at an approximately 50—60 degrees angle. With the bevel facing down, once the needle breaches the joint capsule, the tip can rest on the femoral neck and a small amount of fluid is injected to open the joint. At this point, the needle can be repositioned if needed and the remaining solution should be seen to flow proximally around the femoral head (Fig. 10.1.1D and E). Care must be taken to avoid vessels and nerves in the groin.

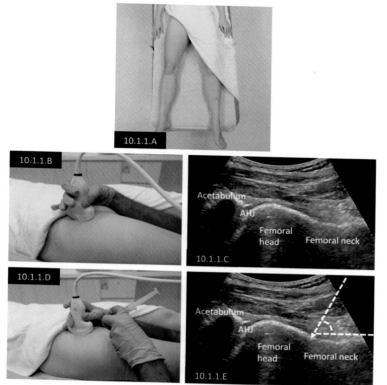

FIGURE 10.1.1 Injections to the AHJ.

In the supine position, the AHJ can be opened further by lowering the end of the couch. The injection can be undertaken using an IP technique from a distal approach, with the transducer in a LAX orientation. The needle tip should be seen within distal joint capsule.

10.1.2 **Pubic symphysis**

Patient position:	For the pubic symphysis (PS), position the patient in a supine position with the abdomen and suprapubic area exposed (Fig. 10.1.2A).
Identifying the anatomy:	The PS can be visualised by placing the transducer in a LAX orientation and following the superior pubic ramus until the symphysis is identified and the transducer lies across the joint (Fig. 10.1.2.B and C). The diarthrodal joint can be easily seen with the pubic rami either side. A SAX view is not usually required for the PS.
Injections performed:	CSI for synovitis, pain symptoms or degenerative disease. PRP injections for pain symptoms or degenerative disease.
Recommended transducer:	Hockey stick 8–18 mHz. Linear 6–15 mHz.
Equipment suggested:	*Equipment preparation:* Set 1 for CSI. Set 4 for PRP injections. *Needle:* 1.5- to 2-inch 23- or 25-gauge needle. *Syringes:* 3 mL for CSI. *Medication:* 40 mg triamcinolone (1 mL) and 1% lidocaine (1 mL). Standard/available PRP preparation.
Injection technique:	Maintaing the transducar in a LAX orientation, the needle can be inserted using an OOP technique, adjacent to the midline of the long edge transducer (Fig. 10.1.2C–E). With the bevel pointing downwards, the needle tip should be seen in the joint before the solution is injected. The solution should be seen to flowing into the PS and there might be tenting of the joint and a pressure sensation for the patient but this should improve with the LA.

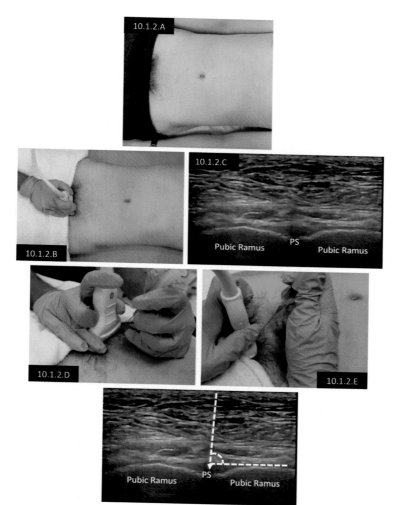

FIGURE 10.1.2 Injections to the PS.

In the supine position, the PS can be seen with the joint in the LAX. The injection can be undertaken using an OOP technique with the needle perpendicular to the skin. The needle tip should be seen within joint itself.

Tendon injections

Superficial tendons commonly injected around the hip area include the gluteus medius (GMT), the iliopsoas (ILT), hamstring origin (HSO) and adductor tendons (AdT).

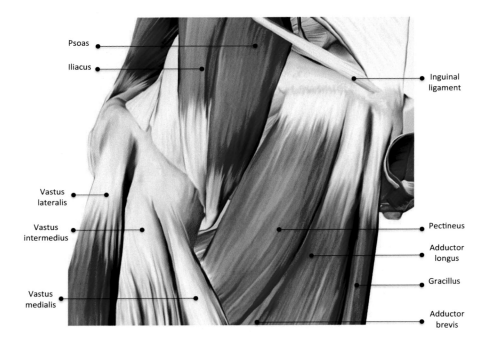

Psoas

Iliacus

Inguinal ligament

Vastus lateralis

Vastus intermedius

Pectineus

Adductor longus

Gracillus

Vastus medialis

Adductor brevis

10.1.3 Gluteus medius tendon

Patient position:	For the gluteus medius tendon (GMT), position the patient in a side-lying position with the asymptomatic contralateral side down on the couch. The symptomatic hip should be exposed and facing up (Fig. 10.1.3A).
Identifying the anatomy:	With the transducer placed in a LAX orientation along the length of the leg, the GMT can be seen inserting on the middle facet of the greater trochanter. Overlying is the tensor fascia lata (TFL) muscle (Fig. 10.1.3B and C), which later becomes the Ilio-Tibial Band. It can also be seen in the SAX by turning the transducer 90 degrees (Fig. 10.1.3D and E).
Injections performed:	CSI for tendinopathy and pain. PRP injections for degenerative tendon disease with intra-substance tearing.
Recommended transducer:	Linear 6–15 mHz. Curvilinear 2–5 mHz.
Equipment suggested:	*Equipment preparation:* Set 1 for CSI. Set 4 for PRP injections. *Needle:* 1.5- to 2-inch 23- or 25-gauge needle. *Syringes:* 5 mL for CSI. *Medication:* 20 mg triamcinolone (0.5 mL) and 1% lidocaine (5 mL). Standard/available PRP preparation.
Injection technique:	Viewing the GMT in the LAX orientation, needle can be introduced IP at approximately 45 degrees (Fig. 10.1.3F and G). Alternatively, the transducer can be turned 90 degrees into an SAX view and the needle can be brought in as horizontal as possible, also utilising an IP technique (Fig. 10.1.3H and I). For CSI, the needle should be aimed at the interface between the GMT and TFL to avoid direct injection into the tendon substance. If a PRP injection is being undertaken, the needle should be advanced into the tendon itself and a fenestration technique can be used to distribute the solution.

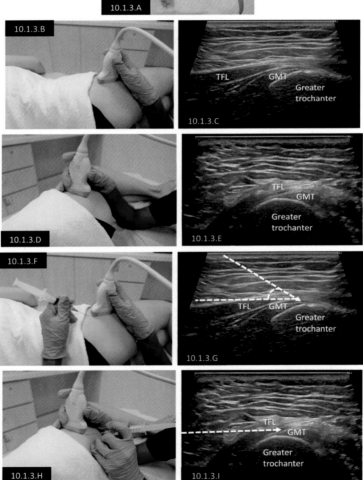

FIGURE 10.1.3 Injections to the GMT.

In the side-lying position, the GMT can be seen in the LAX attaching to the greater trochanter with the TFL overlying. The injection can be undertaken using an IP technique in the SAX with the needle introduced as horizontal as possible to the transducer. The needle tip should be seen against the body of the tendon for CSI or within for PRP.

10.1.4 Iliopsoas tendon

Patient position:	For the iliopsoas tendon (ILT), the patient should be positioned in a supine position with the hip slightly extended by lowering the end of the couch (Fig. 10.1.4A). This exposes the ILT and aims to make it more superficial.
Identifying the anatomy:	With the transducer placed in a LAX orientation over the hip joint, it can be moved medially to identify the ILT as it travels to its attachment at the lesser trochanter. In this position, the vascular bundle can be seen nearby. Once found, the ILT can be traced proximally before the transducer is turned 90 degrees to view the ILT in a SAX view (Fig. 10.1.4B and C).
Injections performed:	CSI for tendinopathy and pain symptoms. PRP injections for degenerative tendon disease with intrasubstance tearing.
Recommended transducer:	Linear 6–15 mHz.
Equipment suggested:	*Equipment preparation:* Set 1 for CSI. Set 4 for PRP injections. *Needle:* 1.5- to 2-inch 23- or 25-gauge needle. *Syringes:* 3 mL for CSI. *Medication:* 20 mg triamcinolone (0.5 mL) and 1% lidocaine (2 mL). Standard/available PRP preparation.
Injection technique:	Viewing the ILT in the SAX, the needle should be introduced at approximately 45 degrees from a lateral approach (Fig. 10.1.4D and E). With the bevel down, it can be aimed at the tendon and once touching against the body of the tendon, a small amount of solution can be injected. Once the tissue planes separate, the needle can be repositioned in this space if required, and the remainder of the solution can be injected. An OOP can also be used with the needle introduced perpendicular to skin and adjacent to the long edge of the transducer (Fig. 10.1.4F and G). If a PRP injection is being undertaken, the needle should be advanced into the tendon itself before a fenestration technique is used to distribute the solution into the area of injury.

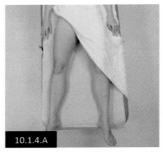

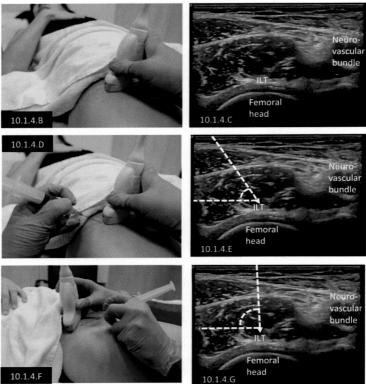

FIGURE 10.1.4 Injections to the ILT.

In the supine position, the hip can be opened further by lowering the end of the couch. The injection can be undertaken using an IP technique from the lateral aspect with the transducer in the SAX or with an OOP approach. The needle tip should be seen against the body of the tendon for CSI or within for PRP.

10.1.5 Adductor tendons

Patient position:	For the adductor tendons (AdT), the patient should be positioned in a supine position with the hip abducted and slightly externally rotated (Fig. 10.1.5A). The knee can be laid to rest on a pillow or rolled towel. This should expose the adductor/groin region. If needed, the lower half of the treatment couch can be lowered to bring the tendons in to a more superficial position.
Identifying the anatomy:	With the transducer placed in a LAX orientation over the adductor muscle, it can be brought to the groin where the conjoint tendon can be seen to attach onto the pubic tubercle (PT). The three adductor muscles (longus, brevis and magnus) are seen (Fig. 10.1.5B and C).
Injections performed:	CSI for tendinopathy and pain. PRP injections for degenerative tendon disease with intrasubstance tearing.
Recommended transducer:	Linear 6–15 mHz.
Equipment suggested:	*Equipment preparation:* Set 1 for CSI. Set 4 for PRP injections. *Needle:* 1.5- to 2-inch 23- or 25-gauge needle. *Syringes:* 3 mL for CSI. *Medication:* 20 mg triamcinolone (0.5 mL) and 1% lidocaine (2 mL). Standard/available PRP preparation.
Injection technique:	Viewing the AdT in a LAX oreintation, the needle can be introduced IP from a distal position at approximately 30 degrees depending on body habitus. With the bevel facing down, the needle can be brought to rest against the tendon and once a small amount of solution has been injected to separate the tissue planes, the needle can be repositioned within this space and the remainder of the solution can be injected (Fig. 10.1.5D and E). If a PRP injection is being undertaken, the needle should be advanced into the tendon itself and using a fenestration technique the solution can be distributed.

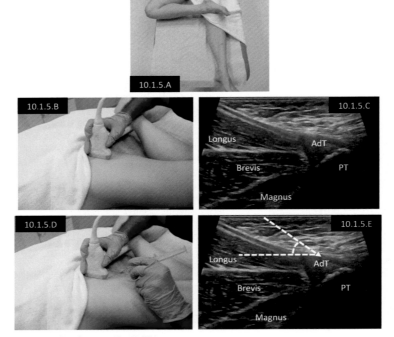

FIGURE 10.1.5 Injections to the AdT.

In the supine position, the hip can be abducted and externally rotated to access the tendons. The injection can be undertaken using an IP technique from a distal approach, with the transducer in a LAX orientation. The needle tip should be seen over the tendons for CSI and within for PRP.

10.1.6 Hamstring origin

Patient position:	For the hamstring origin (HSO), the patient should be positioned in a prone or side-lying position. In the former, the hip can be flexed by lowering the end of the examination couch (Fig. 10.1.6A), and in the latter, the hip and knee can be flexed (Fig. 10.1.6B). Doing so in both positions, brings the HSO to a more superficial position.
Identifying the anatomy:	The transducer can be placed in a LAX orientation over the ischial tuberosity and the HSO can be seen to attach to this (Fig. 10.1.6C and D). Following it in the LAX position, the musculotendinous junction can be seen as it is moved distally (Fig. 10.1.6.E). Over the attachment, the transducer can be rotated 90 degrees (Fig. 10.1.6F and G) to view the tendon in a SAX orientation (Fig. 10.1.6H).
Injections performed:	CSI for tendinopathy or bursitis. PRP injections for degenerative tendon disease with intrasubstance tearing.
Recommended transducer:	Linear 3–12 mHz.
Equipment suggested:	*Equipment preparation:* Set 1 for CSI. Set 4 for PRP injections. *Needle:* 2- to 2.5-inch 21- or 23-gauge needle. *Syringes:* 3 mL for CSI. *Medication:* 20 mg triamcinolone (0.5 mL) and 1% lidocaine (2.5 mL). Standard/available PRP preparation.
Injection technique:	Identifying the HSO in a SAX orientation, with the bevel facing down, the needle can be introduced using an IP technique as horizontal as possible so that it rests upon the tendon ad beneath the gluteus maximus muscle (Fig. 10.1.6I–K). Here a small amount of solution can be injected to separate the tissue planes and open the space. Having done so, the needle can be repositioned if required, before the remaining solution is injected over the attachment. A LAX orientation can also be used to perform the injection but depending on the habitus of the individual it may be more challenging due to the thickness of the gluteal muscle (Fig. 10.1.6L–N). If a PRP injection is being performed, the needle can be inserted into the tendon and an OOP view can be used to confirm placement. A fenestration technique can be used to distribute the solution in the tendon.

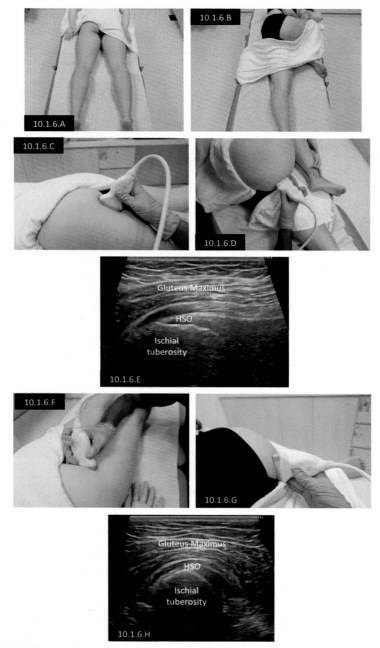

FIGURE 10.1.6 Injections to the HSO.

In the prone or side-lying position, the HSO can be made more superficial by flexing the hip. The injection can be undertaken using an IP technique with the transducer in a LAX or SAX orientation. The needle tip should be seen over the tendons for CSI and within for PRP.

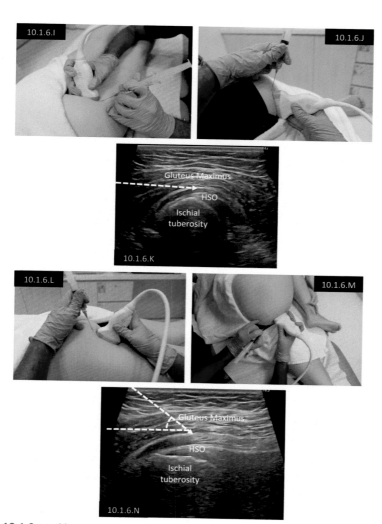

FIGURE 10.1.6 cont'd

Bursa injections

The trochanteric bursa (TB) is situated on the lateral aspect of the hip, between the tensor fascia lata (TFL) and GMT, is a common source of hip pain after minor trauma and can be injected under ultrasound guidance.

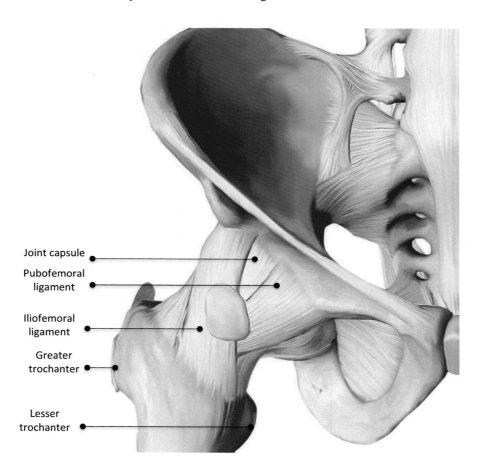

Joint capsule

Pubofemoral ligament

Iliofemoral ligament

Greater trochanter

Lesser trochanter

10.1.7 Trochanteric bursa

Patient position:	For the trochanteric bursa (TB), position the patient in a side-lying position with the symptomatic hip exposed and facing up (Fig. 10.1.7A). If the contralateral hip is asymptomatic, then the patient can rest this upon the treatment couch, whereas if that is also symptomatic, cushioning should be provided.
Identifying the anatomy:	With the transducer placed in a LAX orientation along the length of the leg, the TB can be seen between the tenor fascia lata (TFL) and gluteus medius tendon (GMT) (Fig. 10.1.7B and C). The transducer can then be turned 90 degrees to view the bursa in a SAX view (Fig. 10.1.7D and E).
Injections performed:	CSI for pain and bursitis.
Recommended transducer:	Linear 6—15 mHz. Curvilinear 2—5 mHz.
Equipment suggested:	*Equipment preparation:* Set 1 for CSI. *Needle:* 1.5- to 2-inch 23- or 25-gauge needle. *Syringes:* 5 mL for CSI. *Medication:* 40 mg triamcinolone (0.5 mL) and 1% lidocaine (4 mL).
Injection technique:	Viewing the bursa in a LAX orientation, the needle can be introduced with the bevel facing down with an IP technique at approximately 40 degrees, until the tip is seen between the TFL and GMT (Fig. 10.1.7F and G). In the SAX, the needle should be brought in as horizontal as possible to enable IP visualisation and again, the needle is aimed for the interface between the GMT and TFL (Fig. 10.1.7H and I). In both situations, once a small amount of fluid is injected to open the space, the needle can be repositioned as required and it is important to avoid direct injection into the tendon beneath or muscle above.

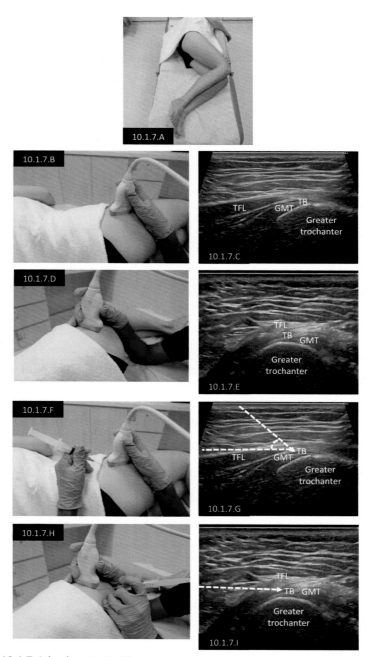

FIGURE 10.1.7 Injections to the TB.

In the side-lying position, the TB can be seen in the LAX between the GMT and TFL. The injection can be undertaken using an IP technique in the SAX with the needle introduced as horizontal as possible to the transducer. The needle tip should be seen between the GMT and TFL.

Nerve injections

The lateral femoral cutaneous nerve (LFCN) is commonly compressed as it passes beneath the inguinal ligament, causing meralgia paraesthetica (pain, tingling and numbness in the anterolateral aspect of the thigh). This can be readily identified and injected under ultrasound guidance.

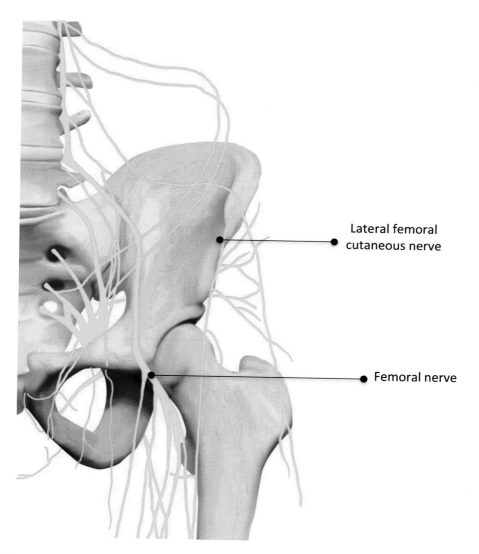

Lateral femoral cutaneous nerve

Femoral nerve

10.1.8 Lateral femoral cutaneous nerve

Patient position:	For the lateral femoral cutaneous nerve (LFCN), the patient should be placed in a supine position with the hip slightly extended by moving the end of the treatment couch downwards (Fig. 10.1.8A). In this position, the patient can be asked to perform a straight leg raise and a hollow between the rectus femoris (RF) and sartorius (SRT) can be seen. This is where the LCFN is located.
Identifying the anatomy:	The transducer can be placed in a SAX oreintation over the anatomical position for the LFCN and it can be seen running in this channel (Fig. 10.1.8B and C). The nerve can be followed proximally to the inguinal ligament where compression normally occurs.
Injections performed:	CSI for meralgia paraesthetica.
Recommended transducer:	Linear 6—15 mHz. Hockey stick 8—18 mHz.
Equipment suggested:	*Equipment preparation:* Set 1 for CSI. *Needle:* 1- to 1.5-inch 25- or 27-gauge needle. *Syringes:* 3 mL for CSI. *Medication:* 20 mg triamcinolone (0.5 mL) and 1% lidocaine (1 mL).
Injection technique:	Viewing the LFCN in the SAX oreintation, the needle can be introduced with the bevel facing down using an IP technique at 10—15 degrees from the lateral aspect so that it rests above the nerve (Fig. 10.1.8D and E). A small amount of solution can be injected to separate the layers and open the space and once the anatomy is more clearly defined, the remainder can be injected around the nerve. Care must be taken not to inject into the nerve directly.

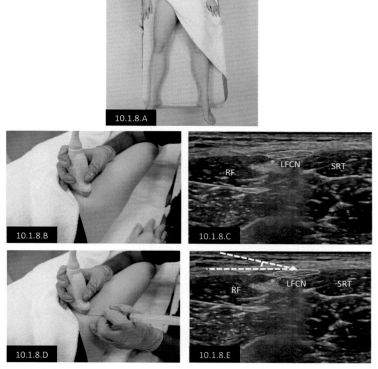

FIGURE 10.1.8 Injections to the LFCN.

In the supine position, the hip can be opened further by lowering the end of the couch. The injection can be undertaken using an IP technique from a lateral approach, with the transducer in a SAX orientation. The needle tip should be seen above the LFCN.

Muscle

Although not commonly undertaken, an intramuscular injection can be performed into the piriformis muscle (PM) for patients experiencing significant radicular pain from sciatic nerve compression as it passes near the PM.

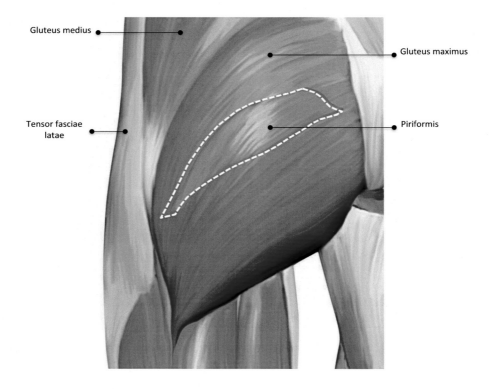

10.1.9 Piriformis muscle

Patient position:	For the piriformis muscle (PM), the patient can be placed in a prone position (Fig. 10.1.9A). To put the area on stretch, the hip can be slightly flexed by lowering the end of the examination couch, thereby taking the PM into a more superficial position.
Identifying the anatomy:	The transducer can be placed in a SAX orientation over the ischial tuberosity and moved proximally and laterally. The PM should then be seen in a LAX view, moving towards the greater trochanter (GT) of the femur beneath the gluteus maximus (GMax) (Fig. 10.1.9B and C). The sciatic nerve can be seen running in close proximity to the PM.
Injections performed:	CSI for pain symptoms.
Recommended transducer:	Linear 6–15 mHz. Curvilinear 2–5mhz.
Equipment suggested:	*Equipment preparation:* Set 1 for CSI. *Needle:* 2- to 2.5-inch 23- or 25-gauge needle. *Syringes:* 5 mL for CSI. *Medication:* 40 mg triamcinolone (0.5 mL) and 1% lidocaine (3 mL).
Injection technique:	Viewing the PM in the LAX orientation, the needle can be introduced using an IP technique at an angle of approximately 45 degrees (Fig. 10.1.9D and E). This may need to be altered according to patient habitus or muscle bulk. With the bevel facing down, the needle tip can be inserted into the muscle and provided there is no blood flow on aspiration, the CSI can be distributed using a fenestration technique. Care should be taken to avoid the sciatic nerve.

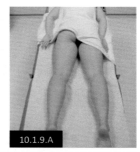

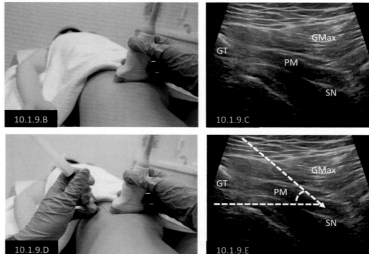

FIGURE 10.1.9 Injections to the PM.

In the prone position, the PM can be made more superficial by flexing the hip. The injection can be undertaken using an IP technique with the transducer in a LAX orientation over the PM. The needle tip should be taken into the PM before the injection is undertaken.

Notes

(Please use this area to reflect on your procedure and how you can build on these experiences).

10.2 **Knee**

With many superficial structures around the knee that can be injured, several can be treated with ultrasound guided interventions. The key with knee injections is stability, sufficient exposure and ease of access to the structure of interest. As such the prone, supine or side-lying position is most suited for knee injections and an appropriate support can help with stability.

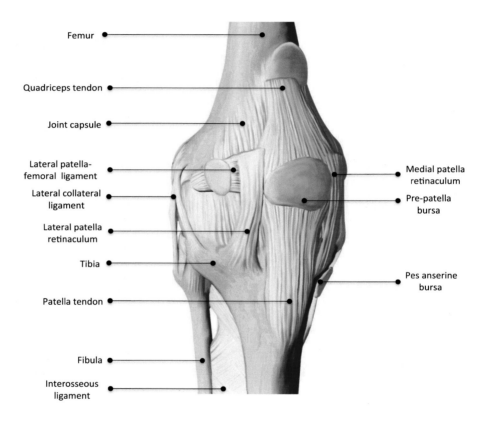

Femur

Quadriceps tendon

Joint capsule

Lateral patella-femoral ligament

Lateral collateral ligament

Lateral patella retinaculum

Tibia

Patella tendon

Fibula

Interosseous ligament

Medial patella retinaculum

Pre-patella bursa

Pes anserine bursa

Joint injections

The common joint injections around the knee include the main knee joint (KJ) itself, proximal tibio-fibular joint (PTJF) and posteriorly into a Baker's cyst (BC).

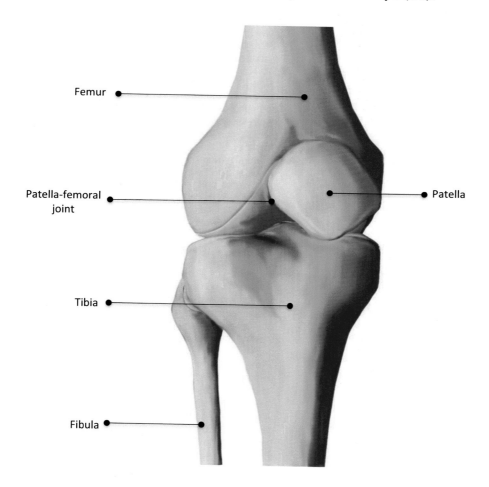

10.2.1 Knee joint

Patient position:	For the knee joint (KJ), the patient can be positioned in a supine position with the knee flexed to approximately 45 degrees. Using a stable support limits movement and eases the guidance of injections (Fig. 10.2.1A and B).
Identifying the anatomy:	The KJ can be accessed via the suprapatellar bursa (SPB) or patello-femoral joint (PFJ). For the former, the transducer is best placed in a SAX orientation over the quadriceps tendon and SPB can be seen beneath it (Fig. 10.2.1C and D). For the PFJ, the transducer can be placed over the medial or lateral border of the patella and medial or lateral femoral condyle (MFC or LFC); the joint space should be seen between the two bony prominences (Fig. 10.2.1E and F).
Injections performed:	CSI for pain or degenerative disease. HA or PRP injections for degenerative disease.
Recommended transducer:	Linear 6—15 mHz (SPB). Hockey stick 8—18 mHz (PFJ).
Equipment suggested:	*Equipment preparation:* Set 1 for CSI and HA injections. Set 4 for PRP injections. *Needle:* 2- to 2.5-inch 21- or 23-gauge needle. *Syringes:* 5 mL for CSI. *Medication:* 40 mg triamcinolone (1 mL) and 1% lidocaine (4 mL). Standard/available HA or PRP preparation.
Injection technique:	Viewing the SPB in a SAX orientation, the needle can be introduced with the bevel down, using an IP technique from the lateral aspect (Fig. 10.2.1G and H). It is often good to start a few centimetres below the transducer so that the needle can be brought in parallel to the field of view. Accessing from the PFJ, the transducer can be placed in a SAX orientation over it and the needle can be introduced using either an IP (Fig. 10.2.1I and J) or OOP technique (Fig. 10.2.1K and L). In the former, the needle may be less visible once in the joint, while with the latter, only the needle tip will be seen in the joint space.

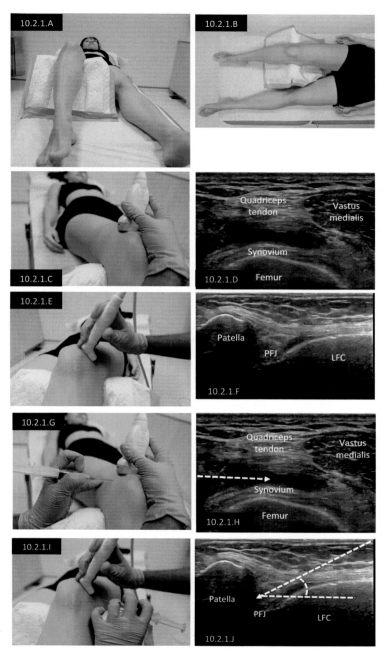

FIGURE 10.2.1 Injections to the KJ.

In the supine position the knee can be flexed to 45 degrees. The injection can be undertaken with the transducer in a SAX orientation over the SPB (IP approach) or PFJ (OOP approach). The needle tip should be seen in the bursa or joint before the injection is undertaken.

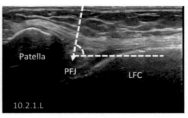

10.2.1.K

10.2.1.L

Patella

PFJ

LFC

FIGURE 10.2.1 cont'd

10.2.2 Proximal tibio-fibular joint

Patient position:	To view the proximal tibio-fibular joint (PTFJ), the patient can be positioned in a supine position with the knee flexed to approximately 45 degrees (Fig. 10.2.2A and B). Using a stable support limits movement and eases injections.
Identifying the anatomy:	The PTFJ can be identified by tracing the tibia laterally in a SAX orientation and the fibula should be seen within the same view. The joint space should be seen between the two bony prominences (Fig. 10.2.1C and D).
Injections performed:	CSI or LA for pain or degenerative disease. PRP injections for degenerative changes.
Recommended transducer:	Linear 6–15 mHz. Hockey stick 8–18 mHz.
Equipment suggested:	*Equipment preparation:* Set 1 for CSI. Set 4 for PRP injections. *Needle:* 2- to 2.5-inch 21- or 23-gauge needle. *Syringes:* 3 mL for CSI. *Medication:* 20 mg triamcinolone (0.5 mL) and 1% lidocaine (1 mL). Standard/available PRP preparation.
Injection technique:	Maintaining the PTFJ in a SAX oreintation, the needle can be introduced perpendicular to the skin, using an OOP technique adjacent to the distal border of the transducer. With the bevel facing down, the needle tip should seen within the joint line (Fig. 10.2.2E and F). Alternatively, an IP technique can be used, approaching from the lateral aspect (Fig. 10.2.1G and H). If this method is used then it is important to avoid injury to the common peroneal nerve or its branches. In either situation once the needle is within the joint, provided there is no blood flow on aspiration, the solution can be injected.

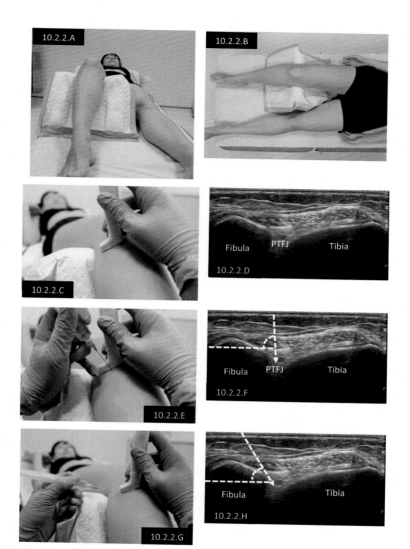

FIGURE 10.2.2 Injections to the PTFJ.

In the supine position the knee can be flexed to 45 degrees. The injection can be undertaken with the transducer in a SAX orientation across the proximal tibia and fibula using an IP or OOP approach. Care should be taken to avoid nearby neurovascular structures.

10.2.3 Baker's cyst

Patient position:	To access a Baker's cyst (BC), the patient can be positioned in a prone position with the knee fully extended to bring the posterior joint line into a superficial position (Fig. 10.2.3A). A rolled towel around the ankle can help stabilise the lower leg and limit movement.
Identifying the anatomy:	The BC can be identified by tracing the medial gastrocnemius proximally or the semi-membranosus (SMT)/semi-tendinosus (STT) tendons distally. As the two structures cross at the posterior-medial border of the knee, the BC should be seen as a hypoechogenic swelling with the neck extending from the joint (Fig. 10.2.3B and C).
Injections performed:	CSI for pain or degenerative disease. HA or PRP injections for degenerative disease.
Recommended transducer:	Linear 6–15 mHz. Curvilinear 2–5 mHz.
Equipment suggested:	*Equipment preparation:* Set 1 for CSI or HA injections. Set 4 for PRP injections. *Needle:* 1.5- to 2-inch 23- or 25-gauge needle. *Syringes:* 5 mL for CSI. 10 mL(s) for drainage. *Medication:* 40 mg triamcinolone (1 mL) and 1% lidocaine (4 mL). Standard/available HA or PRP preparation.
Injection technique:	Maintaining the transducer in a SAX oreintation, the needle can be introduced from the medial aspect of the knee, using an IP technique. With the bevel facing down, the needle can be angled at approximately 30 degrees, but this might need adjustments depending on body habitus. The needle is advanced until the tip is seen within the BC itself (Fig. 10.2.3D and E). It is important to ensure that the neurovascular bundle is visualised and avoided. Provided there is no blood flow on aspiration, the solution can be injected.

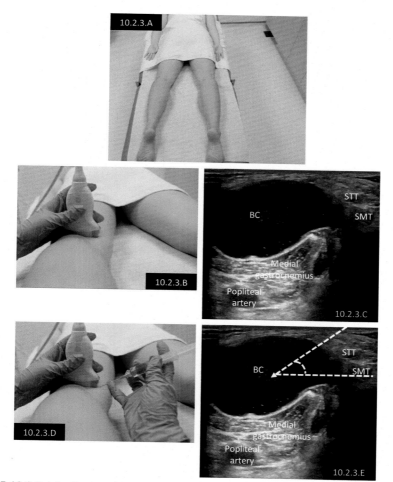

FIGURE 10.2.3 Injections to the BC.

In the prone position, the knee is fully extended, and the injection can be undertaken with the transducer in a SAX orientation across the popliteal fossa. The needle is introduced using an IP technique. Care should be taken to avoid nearby neurovascular structures.

Tendon injections

With multiple tendons around the knee joint, those commonly requiring an injection include the quadriceps (QT), the patella (PT), biceps femoris (BFT) or popliteus tendon (PopT). While it is not recommended to put steroid around weight-bearing tendons such as the QT or PT, clinical judgement may be required depending on the patient's presentation and clinical need.

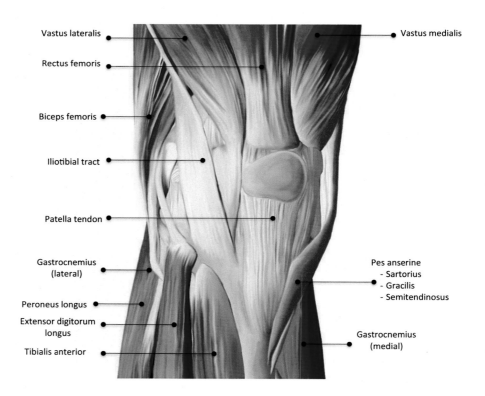

Vastus lateralis

Vastus medialis

Rectus femoris

Biceps femoris

Iliotibial tract

Patella tendon

Gastrocnemius (lateral)

Pes anserine
- Sartorius
- Gracilis
- Semitendinosus

Peroneus longus

Extensor digitorum longus

Gastrocnemius (medial)

Tibialis anterior

10.2.4 Quadriceps tendon

Patient position:	For the quadriceps tendon (QT), the patient can be positioned in a supine position with the knee flexed to approximately 45 degrees (Fig. 10.2.4A and B). Using a stable support places the tendon under a slight tension and limits movement to ease injections.
Identifying the anatomy:	The QT can be visualised by placing the transducer in a LAX orientation to identify the longitudinal structure of the tendon (Fig. 10.2.4C and D). It can be traced proximally to confirm that it is intact. Turning the transducer perpendicular to this position, the tendon can be viewed in a SAX view (Fig. 10.2.4E and F). In both planes, if a joint effusion is present, a prominent bursa may be noted beneath it.
Injections performed:	PRP injections for degenerative tendon disease with intrasubstance tearing.
Recommended transducer:	Linear 6—15 mHz.
Equipment suggested:	*Equipment preparation:* Set 4 for PRP injections. *Needle:* 1.5- to 2-inch 25- or 27-gauge needle.
Injection technique:	Standard/available HA or PRP preparation. Viewing the QT in a SAX orientation, the needle can be introduced with the bevel facing down, using an IP technique from the lateral aspect. The aim is to keep the needle parallel to the transducer (Fig. 10.2.4G and H). Aiming for the area of interest in the tendon, the solution can be injected using a fenestration technique. The position of the needle in the tendon be assessed using an OOP view by rotating the transducer into a LAX orientation. An LAX approach can also be used, with the needle also introduced using an IP technique at an angle of approximately 30 degrees (Fig. 10.2.4I and J). A similar fenestration technique should be used to inject the PRP.

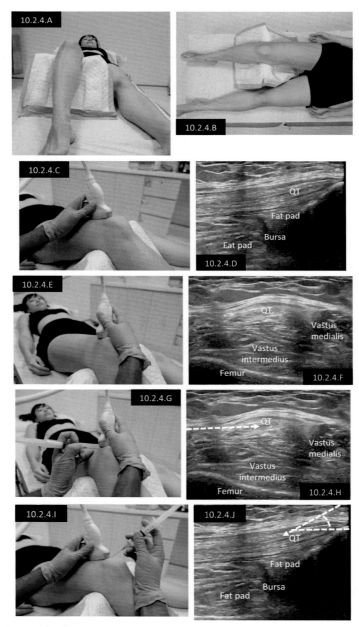

FIGURE 10.2.4 Injections to the QT.

In the supine position the knee can be flexed to 45 degrees. The injection can be undertaken with the transducer in a SAX or LAX orientation over the QT, and the needle is introduced using an IP technique into the tendon. A fenestration technique can be used to inject the PRP.

10.2.5 Patella tendon

Patient position:	For the patella tendon (PT), the patient can be positioned in a supine position with the knee flexed to approximately 45 degrees (Fig. 10.2.5A and B). Using a stable support places the tendon under a slight tension and limits movement during the procedure.
Identifying the anatomy:	The PT can be visualised by placing the transducer in a LAX orientation to identify the longitudinal structure of the tendon and confirming that its proximal (to the distal pole of the patella) and distal (to the tibial tuberosity) attachments are intact. Turning the transducer perpendicular to this position, the PT can be viewed in a SAX view (Fig. 10.2.5C and D). The fat pad should be visualised beneath the tendon.
Injections performed:	HVI tendon stripping for tendinopathy. PRP injections for degenerative tendon disease with intrasubstance tearing.
Recommended transducer:	Linear 6–15 mHz.
Equipment suggested:	*Equipment preparation:* Set 4 for PRP injections. *Syringe:* 10 mL(s) for tendon stripping from the fat pad. *Medication:* 1% lidocaine (5 mL) and normal saline (15–20 mL). *Needle:* 1.5- to 2-inch 25- or 27-gauge needle. Standard/available PRP preparation.
Injection technique:	Viewing the PT in a SAX orientation, the needle can be introduced using an IP technique from the lateral aspect (Fig. 10.2.5E). For tendon stripping, keep the needle parallel to the transducer, and aimed at the posterior border of the tendon. With the bevel facing upward, it is angled upwards to the interface between the PT and fat pad (Fig. 10.2.5F). Local anaesthetic and saline can be injected to separate the two layers and the needle can be repositioned if needed. For PRP injections into the PT, the needle entry can be higher, parallel to the transducer and directly into the body of the tendon (Fig. 10.2.5G). Once the tip is in position, the procedure can be undertaken using a fenestration technique to distribute the PRP. The position of the needle can also be assessed using an OOP view by turning the transducer into a LAX oreintation and observing the position in terms of depth within the tendon.

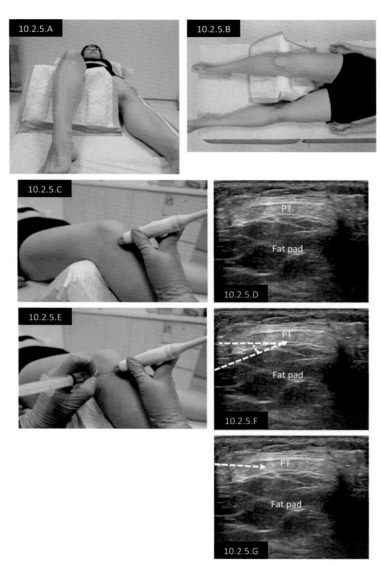

FIGURE 10.2.5 Injections to the PT.

In the supine position the knee can be flexed to 45 degrees. The injection can be undertaken with the transducer in a SAX orientation over the PT, and the needle is introduced using an IP technique either beneath the tendon for a HVI or into it for a PRP injection. A fenestration technique can be used to inject the PRP.

10.2.6 Biceps femoris tendon

Patient position:	To image the biceps femoris tendon (BFT), the patient can be positioned in a prone position with the knee fully extended to bring the posterior joint line into a superficial position (Fig. 10.2.6A). A rolled towel around the ankle can help stabilise the lower leg and limit movement.
Identifying the anatomy:	The BFT can be identified by tracing the biceps femoris muscle distally in a SAX orientation. It can be seen transitioning from the muscle to the tendon. When found, the tendon can be evaluated in the LAX (Fig. 10.2.6B and C) and SAX (Fig. 10.2.6D and E) for any obvious changes.
Injections performed:	CSI for tendinopathy or pain symptoms. PRP injections for degenerative tendon disease with intrasubstance tearing.
Recommended transducer:	Linear 6—15 mHz.
Equipment suggested:	*Equipment preparation:* Set 1 for CSI. Set 4 for PRP injections. *Syringe:* 3 mL for CSI. *Medication:* 20 mg triamcinolone (0.5 mL) and 1% lidocaine (1—2 mL). *Needle:* 1- to 1.5-inch 25- or 27-gauge needle. Standard/available PRP preparation.
Injection technique:	Maintaining the transducer in a LAX orientation, view of the tendon, the needle can be introduced using an IP technique at approximately 20 degrees from a proximal direction (Fig. 10.2.6F and G). With the bevel facing down, the needle tip can be brought to rest against the tendon and a small amount of fluid is injected to separate the tendon sheath. Once separated, the needle can be readjusted if needed and the remaining solution can be pushed through. In the SAX view, the needle can be introduced from the lateral aspect, using an IP technique at 5—10 degrees. Again, the bevel should be facing down and it is advanced so the tip is seen resting on the tendon (Fig. 10.2.6H and I). A small amount of fluid should be injected to separate the tendon sheath before completing the procedure. Either approach can be used for PRP injections with the being that the needle is passed into the tendon. Once within, the PRP is injected using a fenestration technique.

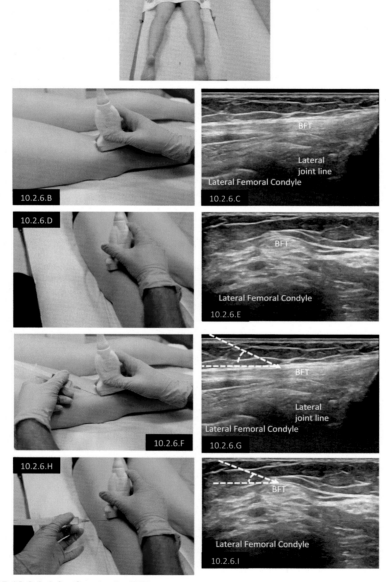

FIGURE 10.2.6 Injections to the BFT.

In the prone position, the knee is fully extended and the injection can be undertaken with the transducer in a SAX or LAX orientation. The needle is introduced using an IP technique between the tendon and its sheath for CSI and into the tendon for PRP injections. A fenestration technique can be used to inject the PRP.

10.2.7 Popliteus tendon

Patient position:	To access a popliteus tendon (PopT), the patient can be positioned in a side-lying position with the knee flexed to 45 degrees and slightly internally rotated (Fig. 10.2.7A). A rolled towel between the legs and also on the examination couch can help stabilise the lower limb and limit movement.
Identifying the anatomy:	The PopT can be found by identifying the lateral joint line in a LAX orientation and then tracing this posteriorly. The tendon can be seen in the SAX within the popliteal groove (Fig. 10.2.7B and C). Turning the transducer perpendicular, the tendon can be then traced in a LAX orientation (Fig. 10.2.7D and E).
Injections performed:	CSI for pain from tendinopathy. PRP injections for degenerative tendon disease with intrasubstance tearing.
Recommended transducer:	Linear 6–15 mHz.
Equipment suggested:	*Equipment preparation:* Set 1 for CSI. Set 4 for PRP injections. *Syringe:* 3 mL for CSI. *Medication:* 20 mg triamcinolone (0.5 mL) and 1% lidocaine (1–2 mL). *Needle:* 1- to 1.5-inch 25- or 27-gauge needle. Standard/available PRP preparation.
Injection technique:	In a SAX view, the needle can be introduced from the lateral aspect of the knee, using an OOP technique and the bevel facing down. It is advanced until the tip is seen to be resting on the tendon (Fig. 10.2.7F and G), after which a small amount of fluid can be injected to separate the sheath. Once separated, the needle can be re-positioned if needed and the remaining solution can be injected. In the LAX view, the needle can be introduced with the bevel facing down, using an IP technique at approximately 15–20 degrees from the lateral aspect of the knee (Fig. 10.2.7H and I). Again, a small amount of fluid is injected to separate the tendon sheath and once a space is noted, the needle can be re-positioned if required, before the remaining solution is injected into this space. Either approach can be used for PRP injections, with the main difference being that the needle is passed into the tendon and the solution is injected using a fenestration technique. Care must be taken to avoid the neurovascular structures around the knee.

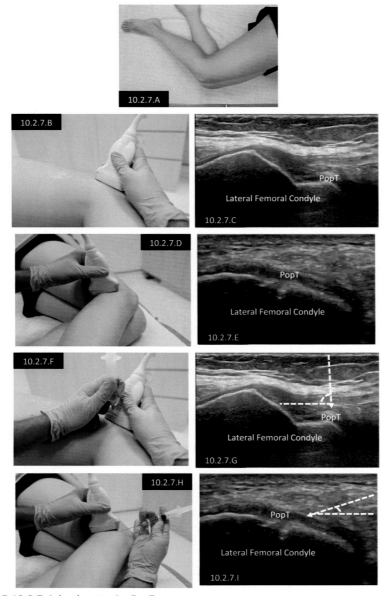

FIGURE 10.2.7 Injections to the PopT.

In the side-lying position, the knee is semi-flexed to 45 degrees, and the injection can be undertaken with the transducer in a SAX (OOP) or LAX (IP) orientation. The needle is introduced between the tendon and its sheath for CSI and into the tendon for PRP injections. A fenestration technique can be used to inject the PRP.

Ligament injections

Ligaments are readily assessed and treated under ultrasound guidance if needed, with the commonest ligament the medial collateral ligament (MCL). Despite being more challenging, the lateral collateral ligament (LCL) is still amenable to a ultrasound guided injections.

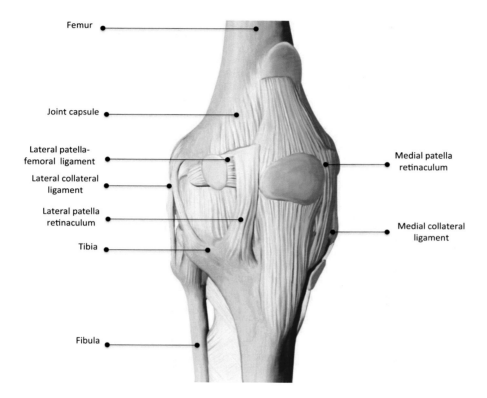

Femur

Joint capsule

Lateral patella-
femoral ligament

Lateral collateral
ligament

Lateral patella
retinaculum

Tibia

Fibula

Medial patella
retinaculum

Medial collateral
ligament

10.2.8 Medial collateral ligament

Patient position:	For the medial collateral ligament (MCL), the patient can be positioned in a supine position with the knee flexed to approximately 45 degrees (Fig. 10.2.8A and B). A stable support can help limit movement and ease injections. The knee can be slightly externally rotated to bring the MCL into prominence.
Identifying the anatomy:	The MCL can be visualised by placing the transducer in a LAX orientation over the medial joint line and then moving it posteriorly until the ligament is seen lying over the joint (Fig. 10.2.8C and D). It can be traced proximally where the ligament is thicker and distally where it narrows down. High resolution imaging can help delineate the deep (dMCL) from superficial MCL (sMCL). The ligament can also be assessed in the SAX by rotating the transducer perpendicular to this orientation.
Injections performed:	CSI for pain symptoms. Prolo injections for degenerative changes or intrasubstance tears.
Recommended transducer:	Linear 6–15 mHz. Hockey stick 8–18 mHz.
Equipment suggested:	*Equipment preparation:* Set 1 for CSI. Set 5 for Prolo. *Needle:* 1- to 1.5-inch 25- or 27-gauge needle. *Syringes:* 3 mL for CSI. *Medication:* 20 mg triamcinolone (1 mL) and 1% lidocaine (1 mL) for CSI. For Prolo, 2 mL of a 50:50 mixture of 50% dextrose and 1% lidocaine.
Injection technique:	Maintaining the transducer in a LAX orientation over the MCL, the needle can be introduced using an IP technique from a proximal orientation. For a CSI, with the bevel facing down, the needles is angled at 15–20 degrees and the the tip is bought to rest either over the superficial or deep MCL (Fig. 10.2.8E and F). A small amount of fluid can be injected to separate the planes, after which the needle position can be readjusted if needed before the remainder is injected. For Prolo injections, the needle tip is brought to rest within the ligament itself and as the solution is injected using a fenestration technique to distribute the solution.

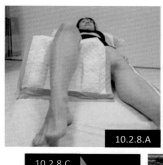
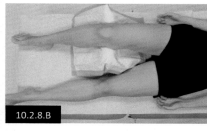

10.2.8.A

10.2.8.B

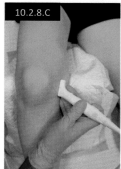

10.2.8.C

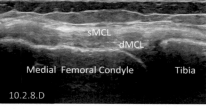

sMCL

dMCL

Medial Femoral Condyle   Tibia

10.2.8.D

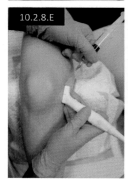

10.2.8.E

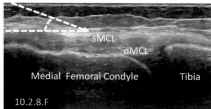

sMCL

dMCL

Medial Femoral Condyle   Tibia

10.2.8.F

FIGURE 10.2.8 Injections to the MCL.

In the supine position the knee can be flexed to 45 degrees. The knee can be slightly externally rotated. The injection can be undertaken with the transducer in a LAX orientation over the MCL, and the needle is introduced using an IP technique either over the ligament for a CSI or into it for a Prolo using a fenestration technique.

10.2.9 Lateral collateral ligament

Patient position:	For the lateral collateral ligament (LCL), the patient can be positioned in a supine position with the knee flexed to approximately 45 degrees (Fig. 10.2.9A and B). A stable support can help limit movement and ease injections and the knee can be placed in a slightly internally rotated position.
Identifying the anatomy:	The LCL can be visualised by placing the transducer in a LAX view over the lateral joint line and then moving it posteriorly until the ligament is seen travelling between the lateral femoral condyle and the fibula (Fig. 10.2.9C and D). The ligament can be quite challenging to find, but can also be assessed in the SAX by rotating the transducer perpendicular to the LAX.
Injections performed:	CSI for pain. Prolo injections for degenerative changes or intrasubstance tears.
Recommended transducer:	Linear 6–15 mHz. Hockey stick 8–18 mHz.
Equipment suggested:	*Equipment preparation:* Set 1 for CSI. Set 5 for Prolo. *Needle:* 1- to 1.5-inch 25- or 27-gauge needle. *Syringes:* 3 mL for CSI. *Medication:* 20 mg triamcinolone (1 mL) and 1% lidocaine (1 mL) for CSI. For Prolo, 2 mL of a 50:50 mixture of 50% dextrose and 1% lidocaine can be used.
Injection technique:	Viewing the LCL in a LAX oreintation, the needle can be introduced using an IP technique from the proximal end with the bevel facing down. Angling at 15–20 degrees, the needle tip is bought to rest either over the LCL (Fig. 10.2.9E and F), at which point a small amount of fluid can be injected to separate the planes. If needed, the needle can be readjusted, after which the remainder can be injected. For Prolo injections, the needle tip is brought to rest within the ligament itself and as the solution is injected, a fenestration technique is used to distribute it within the ligament.

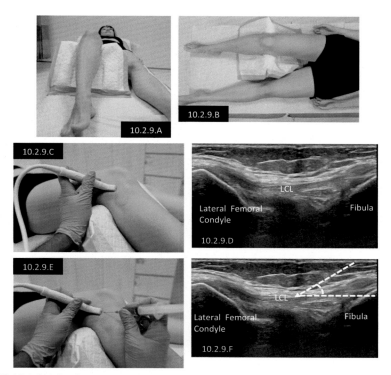

FIGURE 10.2.9 Injections to the LCL.

In the supine position the knee can be flexed to 45 degrees. The knee can also be slightly internally rotated. The injection can be undertaken with the transducer in a LAX orientation over the LCL, and the needle is introduced using an IP technique either over the ligament for a CSI or into it for a Prolo using a fenestration technique.

Bursa injections

Other than the suprapatellar bursa to access the knee joint, common areas amenable to treatment under ultrasound guidance include the pes anserine (PAB), ilio-tibial (ITB) and pre-patellar (PPB) bursas.

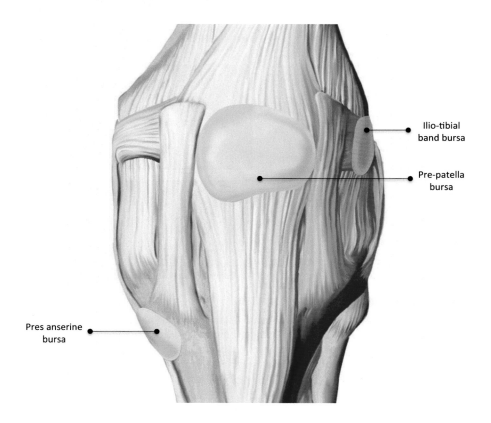

Ilio-tibial band bursa

Pre-patella bursa

Pres anserine bursa

10.2.10 **Pes anserine bursa**

Patient position:	For the pes anserine bursa (PAB), the patient can be positioned in a supine position with the knee flexed to approximately 45 degrees (Fig. 10.2.10A and B). Using a stable support limits movement and eases injections and the knee can be placed in slight external rotation.
Identifying the anatomy:	The PAB can be visualised by placing the transducer in a LAX oreintation over the medial aspect of the knee to identify the pes anserine (made up of the sartorius, semi-tendinosus and gracillus tendons) (Fig. 10.2.10C and D). The bursa will be seen in this area and sometimes can encompass the tendons themselves.
Injections performed:	CSI for pain symptoms.
Recommended transducer:	Linear 6–15 mHz. Hockey stick 8–18 mHz.
Equipment suggested:	*Equipment preparation:* Set 1 for CSI. *Needle:* 1- to 1.5-inch 25- or 27-gauge needle. *Syringes:* 3 mL for CSI. *Medication:* 40 mg triamcinolone (1 mL) and 1% lidocaine (1 mL) for CSI.
Injection technique:	Viewing the PAB in a LAX oreintation, the needle can be introduced using an IP technique from the proximal aspect at an angle of 30–40 degrees (Fig. 10.2.10E and F). With the bevel facing down, the needle should be brought into the bursa or around the tendons and a small amount of solution can be injected. Once flow is seen either into the bursa or around the tendons, the needle can be repositioned if needed, before the remaining solution is injected. Care must be taken not to inject directly into the tendons themselves. An IP SAX approach can also be used to guide the injection, but this can be more challenging to achieve.

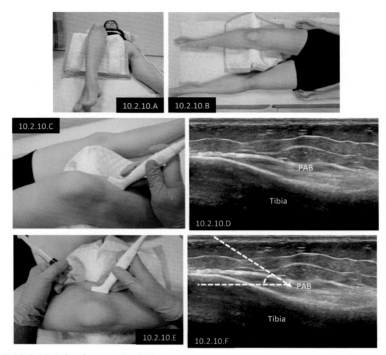

FIGURE 10.2.10 Injections to the PAB.

In the supine position the knee can be flexed to 45 degrees. The knee can also be slightly externally rotated. The injection can be undertaken with the transducer in a LAX orientation over the PAB, and the needle is introduced using an IP technique into the bursa from a distal approach.

10.2.11 Ilio-tibial band bursa

Patient position:	For the ilio-tibial band bursa (ITBB), the patient can be positioned in a supine position with the knee flexed to approximately 45 degrees (Fig. 10.2.11A and B). Using a stable support limits movement and eases injections and the knee can be placed in slight internal rotation to accentuate the lateral structures.
Identifying the anatomy:	The ITBB can be visualised by placing the transducer in a LAX orientation and following the ITB distally until it is seen to cross the lateral femoral condyle (LFC). Here, a hypoechogenic area may be seen between the ITB and LFC. Rotating the transducer perpendicular to this view, the tendon can be viewed in the SAX (Fig. 10.2.11C and D).
Injections performed:	CSI for ilio-tibial band friction syndrome and pain.
Recommended transducer:	Linear 6–15 mHz. Hockey stick 8–18 mHz.
Equipment suggested:	*Equipment preparation:* Set 1 for CSI. *Needle:* 1- to 1.5-inch 25- or 27-gauge needle. *Syringes:* 3 mL for CSI. *Medication:* 40 mg triamcinolone (1 mL) and 1% lidocaine (1 mL) for CSI.
Injection technique:	Viewing the ITBB in the SAX, the needle can be introduced using an IP technique from a lateral approach at an angle of 20–30 degrees with the bevel facing down (Fig. 10.2.11E and F). The needle should be brought into the bursa and a small amount of solution can be injected to open the space, before it is readjusted if needed and the remaining solution is injected. It is important not to inject directly into the ilio-tibial band and nearby neural structures must be avoided.

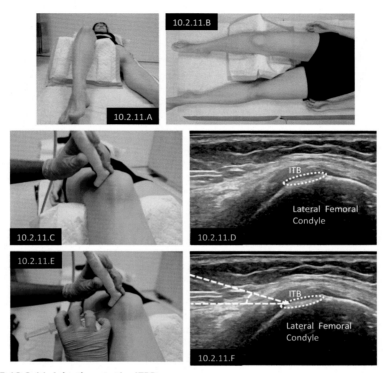

FIGURE 10.2.11 Injections to the ITBB.

In the supine position the knee can be flexed to 45 degrees. The knee can also be slightly internally rotated. The injection can be undertaken with the transducer in a SAX orientation over the ITBB, and the needle is introduced using an IP technique into the bursa. Care must be taken not to damage nearby neurovascular structures.

10.2.12 Pre-patellar bursa

Patient position:	For the pre-patellar bursa (PPB), the patient can be positioned in a supine position with the knee flexed to approximately 45 degrees (Fig. 10.2.12A and B). Using a stable support limits movement and eases injections and the knee can be placed in slight internal rotation.
Identifying the anatomy:	The PPB can be visualised by placing the transducer in a LAX oreintation over the patella and a hypoechogenic area may be seen above the bone (Fig. 10.2.12C and D). Rotating the transducer perpendicular to this to view, the PPB can be viewed in the SAX (Fig. 10.2.12E and F).
Injections performed:	Drainage and possible CSI for pre-patella bursitis or pain.
Recommended transducer:	Linear 6—15 mHz. Hockey stick 8—18 mHz.
Equipment suggested:	*Equipment preparation:* Set 1 for drainage and CSI. *Needle:* 1- to 1.5-inch 25- or 27-gauge needle. *Syringes:* 10 mL for drainage and 3 mL for CSI. *Medication:* 20 mg triamcinolone (1 mL) and 1% lidocaine (1 mL) for CSI.
Injection technique:	Viewing the PPB in the LAX (Fig. 10.2.12G and H) or SAX (Fig. 10.2.12I and J), the needle can be introduced using an IP technique from a proximal or lateral approach at an angle of 10—20 degrees. With the bevel facing down, the needle should be brought into the bursa and the aspiration can be attempted, but sometimes if the bursa has been present for a while, the fluid inside may become viscous. Flushing the space with LA may help loosen some of the viscous fluid and once removed, the syringe can be changed and a CSI is undertaken.

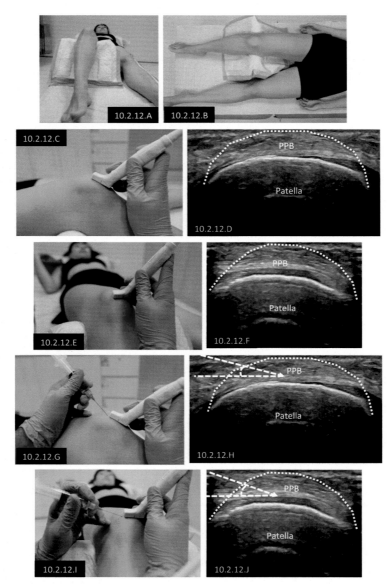

FIGURE 10.2.12 Injections to the PPB.

In the supine position, the knee can be flexed to 45 degrees. The injection can be undertaken with the transducer in a SAX or LAX orientation over the PPB, and the needle is introduced using an IP technique into the bursa. It can be drained before an injection is undertaken.

Muscle

Injury to the quadriceps muscles, particularly tears of the rectus femoris (RF), can be assessed and treated.

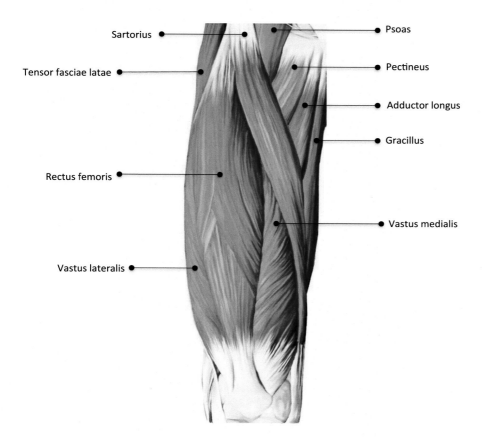

10.2.13 Rectus femoris

Patient position:	To visualise the rectus femoris (RF), the patient can be positioned in a supine position with the knee flexed to approximately 45 degrees (Fig. 10.2.13A). Using a stable support limits movement and eases injections.
Identifying the anatomy:	The RF can be visualized by placing the transducer in a SAX view to identify the quadriceps muscles with the RF being most superficial (Fig. 10.2.13B and C). Tears or disruption to the muscle can be identified by scanning along the length of it and if seen, can be confirmed in a by rotating the transducer perpendicular into a LAX view (Fig. 10.2.13D and E).
Injections performed:	Haematoma aspiration and PRP injections for muscle tears.
Recommended transducer:	Linear 6—15 mHz.
Equipment suggested:	*Equipment preparation:* Set 4 for PRP injections. *Syringe:* 10 mL syringe(s) for aspiration. *Needle:* 1.5- to 2-inch 21- or 23-gauge needle. Standard/available PRP preparation.
Injection technique:	Viewing the RF in the SAX view, the needle can be introduced using an IP technique from the medial or lateral aspect at an angle of 30—40 degrees (Fig. 10.2.13F and G). With the bevel facing down, the needle should be brought into the tear and the haematoma can be aspirated. Once this has been completed, the syringe can be swapped to the PRP-containing one and this can be slowly injected into the space. An LAX approach can also be used to deliver this treatment with the needle introduced from an IP proximal approach at 30—40 degrees (Fig. 10.2.13H and I). With either approach, care must be taken not to damage intact muscle tissue.

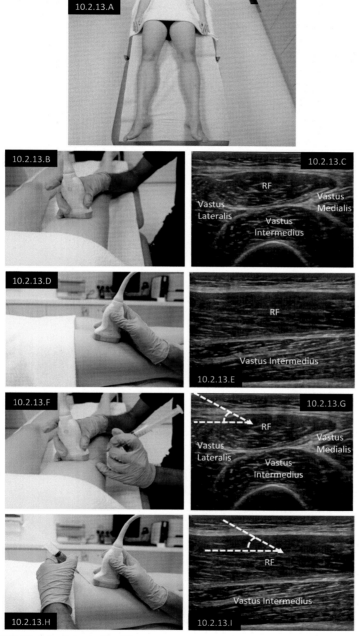

FIGURE 10.2.13 Injections to the RF.

In the supine position the knee can be flexed to 45 degrees, and the injection can be undertaken with the transducer in a SAX or LAX orientation over the RF. The needle is introduced using an IP technique into the tear, and the haematoma can be aspirated before the PRP injection is undertaken.

Notes
(Please use this area to reflect on your procedure and how you can build on these experiences).

10.3 Ankle and foot

With many structures around the ankle and foot being superficial, these are amenable to assessment and treatment under ultrasound guidance. This is particularly useful for areas that may be less easily treated by injections after palpation alone.

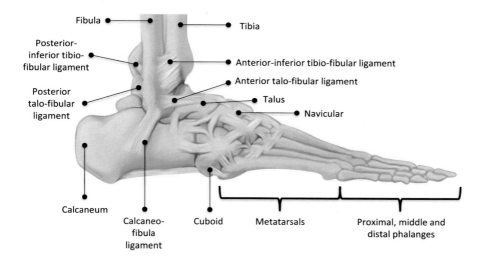

Fibula

Tibia

Posterior-inferior tibio-fibular ligament

Anterior-inferior tibio-fibular ligament

Anterior talo-fibular ligament

Posterior talo-fibular ligament

Talus

Navicular

Calcaneum

Calcaneo-fibula ligament

Cuboid

Metatarsals

Proximal, middle and distal phalanges

Joint injections

Common joints that are injected in the foot and ankle include the talo-crural joint (TCJ), the subtalar joint (STJ), the sinus tarsi (ST), talo-navicular joint (TNJ), metatarso-phalangeal joint (MTPJ) and inter-phalangeal joints (IPJ). With the smaller joints, a more concentrated solution is useful to obtain the therapeutic effect.

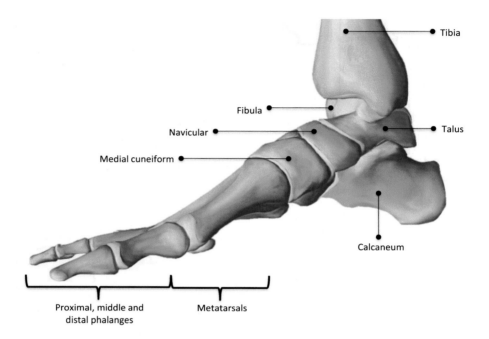

10.3.1 **Talo-crural joint**

Patient position:	For the talo-crural joint (TCJ), the patient can be positioned in a supine position with the knee flexed to approximately 45 degrees and if needed a support beneath the knee can add stability. The foot can be planted on the examination couch and the ankle is in a plantar flexed position (Fig. 10.3.1A). In patients with degenerative disease, extending the knee helps open the ankle joint further, particularly if there are large osteophytes.
Identifying the anatomy:	The TCJ can be imaged over the dorsal aspect of the ankle by placing the transducer in a LAX orientation and following the distal tibia the talar dome and joint are seen within the same field of view (Fig. 10.3.1B and C). The joint capsule might be seen with the anterior fat pad overlying.
Injections performed:	CSI for pain, degenerative disease or synovitis. HA or PRP injections for degenerative disease.
Recommended transducer:	Linear 6–15 mHz. Hockey stick 8–18 mHz.
Equipment suggested:	*Equipment preparation:* Set 1 for CSI or HA injections. Set 4 for PRP injections. *Syringe:* 3 mL for CSI. *Medication:* 40 mg triamcinolone (1 mL) and 1% lidocaine (1–2 mL). *Needle:* 1- to 1.5-inch 25- or 27-gauge needle. Standard/available HA or PRP preparation.
Injection technique:	Viewing the TCJ in the LAX orientation, the needle can be introduced using an IP technique from the distal aspect at an angle of approximately 30 degrees. With the bevel facing down, the tip is aimed for the joint (Fig. 10.3.1D and E). Once inside, and provided there is no blood flow on aspiration, the solution can be injected and should be seen flowing into the joint. Alternatively an OOP approach can be employed, with the needle perpendicular to the skin at the midpoint of the long border of the tansducer, which is held in a LAX oreintation (Fig. 10.3.1F and G). Care must be taken to avoid injury to overlying neurovascular structures and tendons.

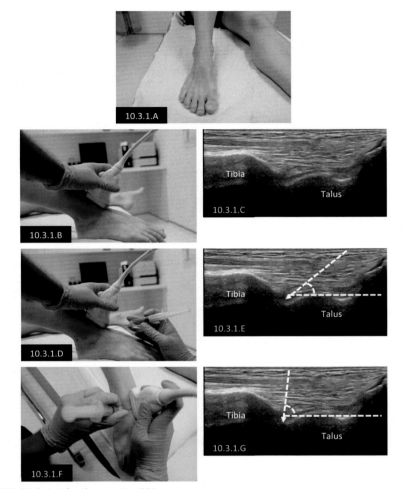

FIGURE 10.3.1 Injections to the TCJ.

In the supine position the knee can be flexed to 45 degrees, and the foot is planted on the couch. The injection can be undertaken with the transducer in a LAX orientation over the TCJ, and the needle is introduced using an IP or OOP technique into joint itself. Care must be taken to avoid injury to the overlying tendons or neurovascular structures.

10.3.2 Subtalar joint

Patient position:	For the subtalar joint (STJ), the patient can be positioned in a supine position with the hip externally rotated and flexed to 45 degrees. The knee is flexed to 90 degrees and the foot is rested on the lateral aspect, with the medial side of the ankle and foot exposed (Fig. 10.3.2A). If needed, a towel or support can be placed under the lateral malleolus to keep the foot in an everted position and open the STJ further.
Identifying the anatomy:	The STJ can be imaged by following the medial malleolus distally in the LAX and identifying the joint as the transducer is moved distally. The joint will be viewed in the SAX as the gap between the medial talus and calcaneum (Fig. 10.3.2.B and C).
Injections performed:	CSI for pain, degenerative disease and synovitis. HA or PRP for degenerative disease.
Recommended transducer	Linear 6—15 mHz. Hockey stick 8—18 mHz.
Equipment suggested:	*Equipment preparation:* Set 1 for CSI or HA injections. Set 4 for PRP injections. *Syringe:* 3 mL for CSI. *Medication:* 20 mg triamcinolone (0.5 mL) and 1% lidocaine (1—2 mL). *Needle:* 1- to 1.5-inch 25- or 27-gauge needle. Standard/available HA or PRP preparation.
Injection technique:	Viewing the STJ in the SAX orientation, the needle can be introduced using an OOP technique by positioning the needle perpendicular to the skin surface midway along the long border of the transducer (Fig. 10.3.2D and E). Once the needle tip is seen within the joint and provided there is no blood flow on aspiration, the solution can be injected and seen to flow into the joint. An IP approach, with the bevel facing down, can also be used from over the calcaneum at an angle of approximately 30 degrees. Care must be taken to identify a gap where neurovascular structures and tendons are missed and hence not injured during the procedure.

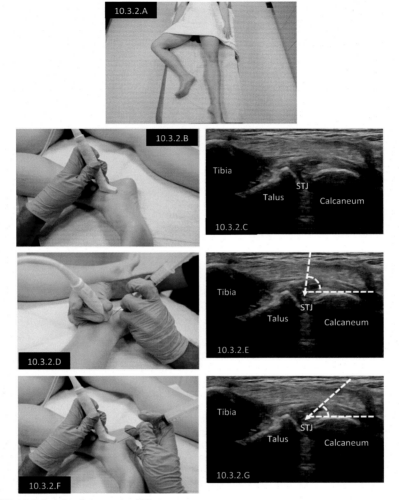

FIGURE 10.3.2 Injections to the STJ.

In the supine position the hip can be abducted, externally rotated, and the foot is resting on the lateral aspect. The injection can be undertaken with the transducer in a LAX orientation over the STJ, which is seen in the SAX, and the needle is introduced using an IP or OOP technique into joint itself. Care must be taken to avoid injury to the overlying neurovascular structures.

10.3.3 **Sinus tarsi**

Patient position:	For the sinus tarsi (ST), the patient can be positioned in a supine position with the knee flexed to approximately 45 degrees and stabilised with a support if needed. The foot can be planted on the examination couch and the ankle is in a plantar flexed, slightly inverted position to open up the space (Fig. 10.3.3A).
Identifying the anatomy:	The ST can be imaged from the dorsal aspect by placing the transducer in a SAX orientation over the lateral malleolus and moving the transducer forward and inferiorly. The depression between the talus and the calcaneum can then be identified (Fig. 10.3.3B and C).
Injections performed:	CSI for pain or degenerative disease. HA or PRP injections for degenerative disease (this should be on the next line).
Recommended transducer:	Linear 6–15 mHz. Hockey stick 8–18 mHz.
Equipment suggested:	*Equipment preparation:* Set 1 for CSI or HA injections. Set 4 for PRP injections. *Syringe:* 3 mL for CSI. *Medication:* 20 mg triamcinolone (0.5 mL) and 1% lidocaine (1–2 mL). *Needle:* 1- to 1.5-inch 25- or 27-gauge needle. Standard/available HA or PRP preparation.
Injection technique:	Viewing the ST in the SAX, the needle can be introduced using an OOP technique adjacent to the long edge of the transducer and perpendicular to the surface of the skin (Fig. 10.3.3D and E). The needle tip should be seen inside the ST, and provided there is no blood flow on aspiration, the solution can be injected. An alternative approach is to introduce the needle using an IP technique from the distal end and coming over the Calcaneum (Fig. 10.3.3F and G). However, this can be more challenging to obtain the appropriate angle and depth.

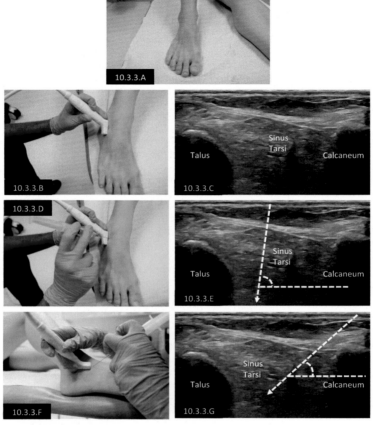

FIGURE 10.3.3 Injections to the ST.

In the supine position the knee can be flexed to 45 degrees, and the foot is planted on the couch. The foot can be also slightly internally rotated to open the lateral aspect. The injection can be undertaken with the transducer in a LAX orientation over the ST, and the needle is introduced using an IP or OOP technique into joint itself. Care must be taken to avoid injury to neurovascular structures.

10.3.4 Talo-navicular joint

Patient position:	For the talo-navicular joint (TNJ), the patient can be positioned in a supine position with the knee flexed to approximately 45 degrees and this can be stabilised with a support if needed. The foot can be planted on the examination couch and the ankle is in a plantar flexed position (Fig. 10.3.4A).
Identifying the anatomy:	The TNJ can be imaged from the dorsal aspect by placing the transducer in a LAX orientation over the ankle joint and following distally. Once the talar neck is identified, the TNJ will be seen as the cortical break between the talus and the navicular bones (Fig. 10.3.4B and C).
Injections performed:	CSI for pain, degenerative disease or synovitis. HA or PRP injections for degenerative disease.
Recommended transducer:	Linear 6—15 mHz. Hockey stick 8—18 mHz.
Equipment suggested:	*Equipment preparation:* Set 1 for CSI or HA injections. Set 4 for PRP injections. *Syringe:* 3 mL for CSI. *Medication:* 20 mg triamcinolone (0.5 mL) and 1% lidocaine (1—2 mL). *Needle:* 1- to 1.5-inch 25- or 27-gauge needle. Standard/available HA or PRP preparation.
Injection technique:	Viewing the TNJ with the transducer in a LAX orientation, the needle can be introduced using an OOP technique with the tip perpendicular to the skin and adjacent to the long border of the probe (Fig. 10.3.4D and E). Once the tip is seem within the joint, provided there is no blood on aspiration, the solution can be injected and there will be flow within the joint. An IP technique in the LAX can be used but may be more challenging if there are significant degenerative changes and osteophytic lipping (Fig. 10.3.4F and G). Care must be taken to avoid injury to neurovascular structures and tendons that may overly the joint.

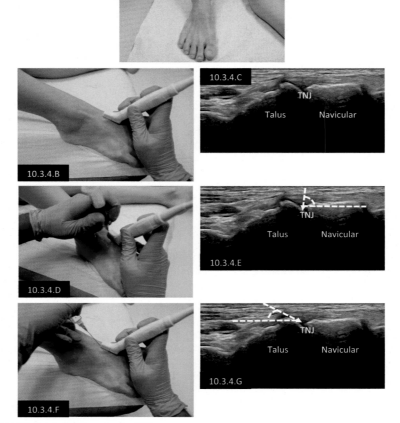

FIGURE 10.3.4 Injections to the TNJ.

In the supine position the knee can be flexed to 45 degrees, and the foot is planted on the couch. The injection can be undertaken with the transducer in a LAX orientation over the TNJ and joint is seen in a SAX view. The needle is introduced using an IP or OOP technique into joint itself. Care must be taken to avoid injury to neurovascular structures overlying.

10.3.5 Metatarso-phalangeal or proximal inter-phalangeal joints

Patient position:	For the metatarso-phalangeal or proximal inter-phalangeal joints (MTPJ/PIPJ), the patient can be positioned in a supine position with the knee flexed to approximately 45 degrees and stabilised with a support if needed. The foot can be planted on the examination couch and the ankle is in a plantar flexed position (Fig. 10.3.5A).
Identifying the anatomy:	The MTP can be imaged from the dorsal aspect by placing the transducer in a LAX orientation and identifying the distal metatarsal and the phalanx in the same plane (Fig. 10.3.5B and C). The discontinuity represents the joint. Similarly, the PIPJ can be seen as a discontinuity between the phalanxes. In degenerative joints, this space may be considerably narrowed by osteophytes and there may be synovial thickening.
Injections performed:	CSI for pain, degenerative disease and synovitis. HA or PRP injections for degenerative disease.
Recommended transducer:	Linear 6–15 mHz. Hockey stick 8–18 mHz.
Equipment suggested:	*Equipment preparation:* Set 1 for CSI or HA injections. Set 4 for PRP injections. *Syringe:* 3 mL for CSI. *Medication:* 20 mg triamcinolone (0.5 mL) and 1% lidocaine (1 mL). *Needle:* 1- to 1.5-inch 25- or 27-gauge needle. Standard/available HA or PRP preparation.
Injection technique:	Viewing the MTPJ/PIPJ with the transducer in a LAX orientation, the needle can be introduced using an OOP technique with the needle perpendicular to the skin surface and adjacent to the long border of the probe. The tip guided into the joint (Fig. 10.3.5D and E) and provided there is no blood on aspiration, the solution can be injected. An IP approach at 15–20 degrees can be used, and it is best to approach from the proximal end with the bevel facing down. The needle is placed against the head of the MT and the solution should be seen to flow into the capsule distally (Fig. 10.3.5F and G). Care must be taken to identify and avoid injury to tendons and neurovascular structures.

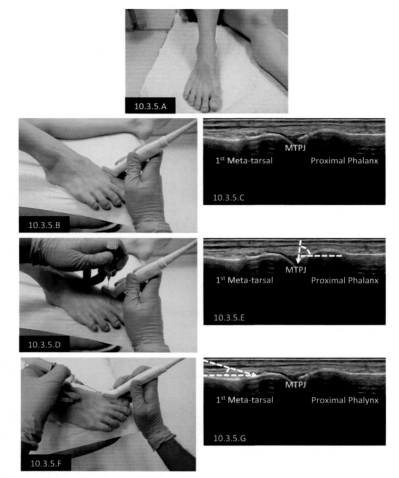

FIGURE 10.3.5 Injections to the MTPJ or PIPJ.

In the supine position the knee can be flexed to 45 degrees, and the foot is planted on the couch. The injection can be undertaken with the transducer in a LAX orientation over the MTPJ or PIPJ, and joint is seen in a SAX view. The needle is introduced using an IP or OOP technique into joint itself. Care must be taken to avoid injury to tendons overlying.

Tendon injections

Tendons commonly injected around the ankle and foot include the peroneus brevis or peroneus longus (PB/PL), the tibialis posterior (TP), the Achilles tendon (AT) and less commonly the flexor hallucis longus (FHL) or tibialis anterior (TA). As with all tendon injections it is important not to inject directly into it with a CSI.

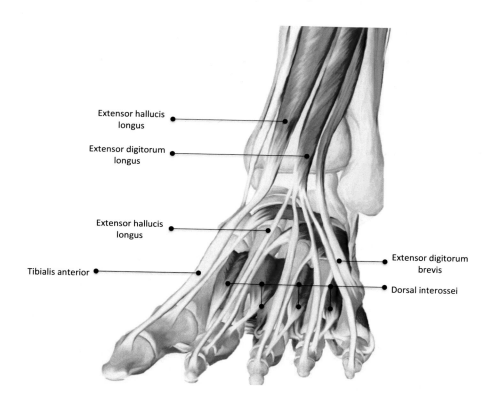

10.3.6 Peroneus longus/peroneus brevis

Patient position:	For the peroneus longus/peroneus brevis (PL/PB) tendons, the patient can be positioned in a prone position with the leg fully extended and the foot hanging off the edge of the bed (Fig. 10.3.6A and B). This exposes the tendons at the posterolateral aspect of the ankle. If needed a towel can be positioned between the dorsum of the foot and treatment couch to add stability.
Identifying the anatomy:	The PL/PB tendons can be traced in the SAX by following the muscles over the posterolateral aspect of ankle with the fibula in view (Fig. 10.3.6C and D). The transition to the tendons will be evident and the tendons can be followed to their attachment in the foot. They can also be assessed in the LAX view although the two tendons might not be quite so distinct (Fig. 10.3.6E and F).
Injections performed:	CSI for acute pain from tendinopathy or tenosynovitis. PRP injections for degenerative tendon disease with intrasubstance tearing.
Recommended transducer:	Linear 6–15 mHz. Hockey stick 8–18 mHz.
Equipment suggested:	*Equipment preparation:* Set 1 for CSI. Set 4 for PRP injections. *Syringe:* 3 mL for CSI. *Medication:* 20 mg triamcinolone (0.5 mL) and 1% lidocaine (1–2 mL). *Needle:* 1- to 1.5-inch 25- or 27-gauge needle. Standard/available PRP preparation.
Injection technique:	Viewing the PL/PB tendons in the SAX orientation, the needle can be introduced using an IP technique from the lateral aspect with the bevel facing down. It should be as horizontal to the transducer as possible (Fig. 10.3.6G and H). Once the needle is in the sheath and rests against the tendon, a small amount of fluid can be injected to separate the tendon from its sheath, before repositioning it if needed, before injecting the remainder. An alternative approach is to introduce the needle using an IP technique in the LAX orientation. Here the angle of approach is approximately 20–30 degrees and again the same technique for injection can be undertaken (Fig. 10.3.6I and J). For PRP injections, the needle direction angle may need to be slightly steeper so that it can be taken into the tendon and a fenestration technique is used to distribute the solution.

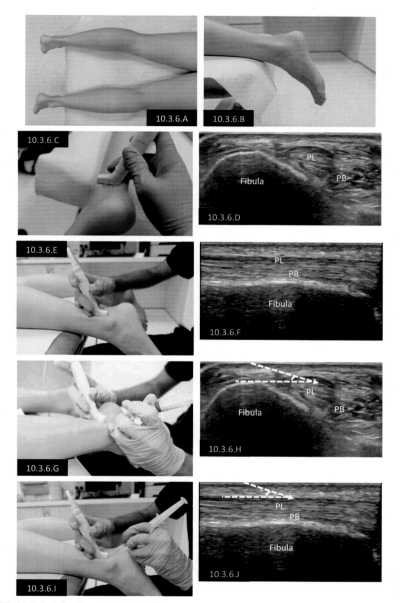

FIGURE 10.3.6 Injections to the PL/PB.

In the prone position the knee is fully extended and foot hangs off the end of the couch. The injection can be undertaken with the transducer in a SAX or LAX orientation over the PL/PB, and the needle is introduced using an IP technique between the tendon and the sheath for CSI and into the tendon itself for PRP injections.

10.3.7 **Tibialis posterior**

Patient position:	For the tibialis posterior (TP), the patient can be positioned in a prone position with the leg fully extended and the foot hanging off the edge of the bed (Fig. 10.3.7A and B). This exposes the posteromedial structures in the ankle. If needed a towel can be positioned between the dorsum of the foot and treatment couch to add stability.
Identifying the anatomy:	The TP can be traced by following the muscles over the posteromedial aspect of ankle with the tibia in the SAX view (Fig. 10.3.7C and D). The transition to the tendons will be evident with the TP most anterior. The tendon can be followed to their attachment in the foot. In the SAX view, the flexor digitorum longus (FDL) and neurovascular bundle can also be seen. Rotating the transducer 90 degrees, the tendon can be viewed in the LAX orientation (Fig. 10.3.7E and F).
Injections performed:	CSI for pain from tendinopathy or tenosynovitis. PRP injections for degenerative tendon disease with intrasubstance tearing.
Recommended transducer:	Linear 6–15 mHz. Hockey stick 8–18 mHz.
Equipment suggested:	*Equipment preparation:* Set 1 for CSI. Set 4 for PRP injections. *Syringe:* 3 mL for CSI. *Medication:* 20 mg triamcinolone (0.5 mL) and 1% lidocaine (1–2 mL). *Needle:* 1- to 1.5-inch 25- or 27-gauge needle. Standard/available PRP preparation.
Injection technique:	Viewing the TP in a SAX orientation, the needle can be introduced with the bevel facing down and using an IP technique from the medial aspect of the ankle as horizontal as possible (Fig. 10.3.7G and H). It should be aimed for the interface between the tendon and sheath; once the tip is in this space, a small amount of fluid can be injected to separate the tissue planes before the needle is re-positioned if needed, before the remainder is injected. An alternative approach is to introduce the needle IP in the LAX orientation. Here the angle of approach is approximately 30 degrees and again the same technique for injection can be undertaken (Fig. 10.3.7I and J). For PRP injections, either approach can be used but the needle is taken into the tendon and a fenestration technique is used to distribute the solution.

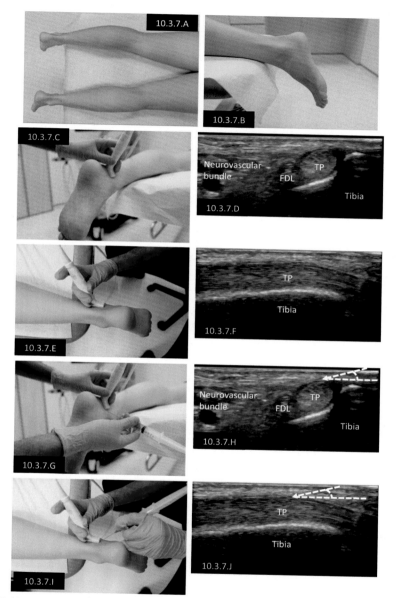

FIGURE 10.3.7 Injections to the TP.

In the prone position the knee is fully extended and the foot hangs off the end of the couch. The injection can be undertaken with the transducer in a SAX or LAX orientation over the TP, and the needle is introduced using an IP technique between the tendon and the sheath for CSI and into the tendon itself for PRP injections.

10.3.8 Achilles tendon

Patient position:	For the Achilles tendon (AT), the patient can be positioned in a prone position with the leg fully extended and the foot hanging off the edge of the bed (Fig. 10.3.8A and B). This exposes the tendons at the posterior aspect of the ankle and if needed, a towel can be positioned between the dorsum of the foot and treatment couch to add stability.
Identifying the anatomy:	The AT tendons can be traced in a LAX orientation, by following the gastrocnemius and soleus muscles at the posterior of the calf, down to the musculartendinous junction and then tendon itself as it attaches on to the calcaneum (Fig. 10.3.8C and D). Rotating the transducer 90 degrees the tendon can be assessed in the SAX orientation as well (Fig. 10.3.8E and F).
Injections performed:	HVI tendon stripping for tendinopathy. PRP injections for degenerative tendon disease with intrasubstance tearing. Prolo injections for retrocalcaneal bursitis.
Recommended transducer:	Linear 6–15 mHz. Hockey stick 8–18 mHz.
Equipment suggested:	*Equipment preparation:* Set 2 for HVI tendon stripping. Set 4 for PRP injections. Set 5 for Prolo. *Syringe:* 10 mL(s) for tendon stripping injections. 3 mL for Prolo injections. *Medication:* High volume tendon stripping normal saline and 1% lidocaine. Prolo, 2 mL of a 50:50 mixture of 50% dextrose and 1% lidocaine can be used. *Needle:* 1- to 1.5-inch 25- or 27-gauge needle. Standard/available PRP preparation.
Injection technique:	In the SAX orientation, the needle can be introduced at 15–20 degrees using an IP technique from the lateral aspect of the ankle. With the bevel facing up, the needle should be aimed for the interface between the tendon sheath and Kager's fat pad (Fig. 10.3.8I and J). Once in position, a small amount of fluid can be injected to separate the planes and the needle can be re-positioned for the HVI. An OOP view helps define the needle position further. While the solution is being pushed through, the needle can be gently moved proximal and distal to distribute it. If vascularity has been particularly florid, the anterior border of the AT can be gently scraped with the needle tip. For PRP injections, the AT can be viewed in the SAX and the needle can be introduced IP from the dorsal aspect at 15–20 degrees directly into the tendon (Fig. 10.3.8G and H). Alternatively, an IP SAX approach can also be used with the needle again guided into the tendon itself (Fig. 10.3.8I and K). In both positions, the solution can be delivered using a fenestration technique. For prolo injections into the retrocalcaneal bursa, a SAX orientation can be used with the needle inserted IP into the space between the tendon and calcaneum where the bursa is found.

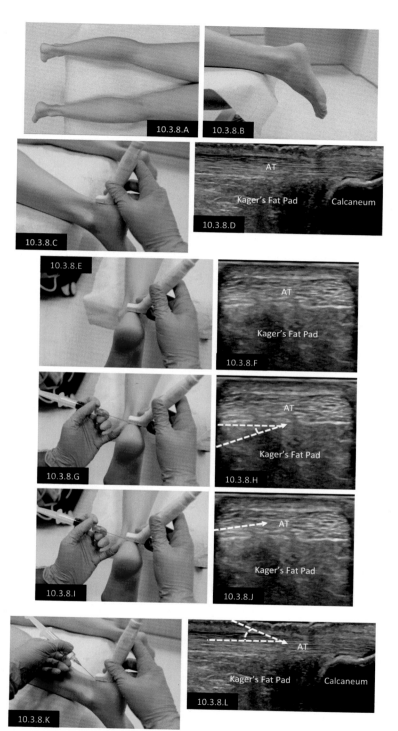

FIGURE 10.3.8 Injections to the AT.

In the prone position the knee is fully extended and the foot hangs off the end of the couch. The injection can be undertaken with the transducer in a SAX or LAX orientation over the AT, and the needle is introduced using an IP technique. For HVI, the needle is placed between the tendon and Kager's fat pad while for PRP injections it is taken into the tendon itself.

10.3.9 Flexor hallucis longus

Patient position:	For the flexor hallucis longus (FHL) tendons, the patient can be positioned in a prone position with the leg fully extended and the foot hanging off the edge of the bed (Fig. 10.3.9A and B). This exposes the tendons at the posteromedial aspect of the ankle and the plantar aspect of the foot. If needed a towel can be positioned between the dorsum of the foot and treatment couch to add stability.
Identifying the anatomy:	The FHL tendons can be traced by following the muscles over the posteromedial aspect of ankle with the tibia in view. The FHL is the deepest of the medial tendons and sits close to the neurovascular bundle. The tendon can be followed around the medial malleolus into the foot and along the plantar surface of the hallux. It is in this distal position that it is more readily visualized sitting between the medial and lateral sessamoids (Fig. 10.3.9C and D). Rotating the transducer 90°, the tendon can also be assessed in a LAX view (Fig. 10.3.9.E and F).
Injections performed:	CSI for pain from tendinopathy or tenosynovitis. PRP injections for degenerative tendon disease with intrasubstance tearing.
Recommended transducer:	Linear 6–15 mHz. Hockey stick 8–18 mHz.
Equipment suggested:	*Equipment preparation:* Set 1 for CSI. Set 4 for PRP injections. *Syringe:* 3 mL for CSI. *Medication:* 20 mg triamcinolone (0.5 mL) and 1% lidocaine (1 mL) for CSI. *Needle:* 1- to 1.5-inch 25- or 27-gauge needle. Standard/available PRP preparation.
Injection technique:	The FHL tends to be deep at the level of the medial malleolus and as such, it may be more appropriate to inject around the tendon at a more distal position between the sessamoid bones. With the bevel facing down, An IP SAX approach can be used for the distal FHL and the needle is directed to between the tendon and sheath at approximately 15–20 degrees (Fig. 10.3.9C and D). Once the tip is in this space and rests against the tendon, a small amount of fluid can be injected to separate the layers before the needle is re-positioned if needed and the remainder is injected. Similarly, an IP LAX approach can be used, with the needle again at 15–20 degrees from a proximal direction (Fig. 10.3.9E and F), however, this may be more painful as more tissue will be cut through. For PRP injections, either approach can be used, with the needle is taken into the tendon itself and a fenestration technique can be used to distribute the solution.

FIGURE 10.3.9 Injections to the FHL.

In the prone position the knee is fully extended and the foot hangs off the end of the couch. The injection can be undertaken with the transducer in a SAX or LAX orientation over the FHL, and the needle is introduced using an IP technique between the tendon and the sheath for CSI and into the tendon itself for PRP injections.

10.3.10 Tibialis anterior

Patient position:	For the tibialis anterior (TA), the patient can be positioned in a supine position with the knee flexed to approximately 45° and this can be stabilised with a support if needed. The foot should be planted flat on the surface of the treatment couch and the ankle is in a plantar flexed position (Fig. 10.3.10A).
Identifying the anatomy:	The TA can be imaged from the dorsal aspect by placing the transducer in a SAX orientation over the ankle and identifying the extensor tendons. The most medial is the TA (Fig. 10.3.10B and C). Once identified, the transducer can be rotated 90 degrees to evaluate the tendon in a LAX view (Fig. 10.3.10D and E).
Injections performed:	CSI for pain from tendinopathy or tenosynovitis. PRP injections for tendinopathy with degenerative tears.
Recommended transducer:	Linear 6—15 mHz. Hockey stick 8—18 mHz.
Equipment suggested:	*Equipment preparation:* Set 1 for CSI. Set 4 for PRP injections. *Syringe:* 3 mL for CSI. *Medication:* 20 mg triamcinolone (0.5 mL) and 1% lidocaine (1 mL) for CSI. *Needle:* 1- to 1.5-inch 25- or 27-gauge needle. Standard/available PRP preparation.
Injection technique:	Viewing the TA in the SAX orientation, the needle can be introduced using an IP technique from a medial or lateral approach at an angle of approximately 10—20 degrees. With the bevel facing down, and the tip aimed for the tendon sheath, (Fig. 10.3.10F and G), once it rests against the tendon, a small amount of fluid can be injected to separate the tissue layers, before the needle is re-positioned and the remainder is injected. An LAX approach can also be used with the needle introduced from the distal or proximal aspect and similar technique is applied (Fig. 10.3.10H and I). For PRP injections, the needle is advanced further into the tendon and the solution is injected using a fenestration technique.

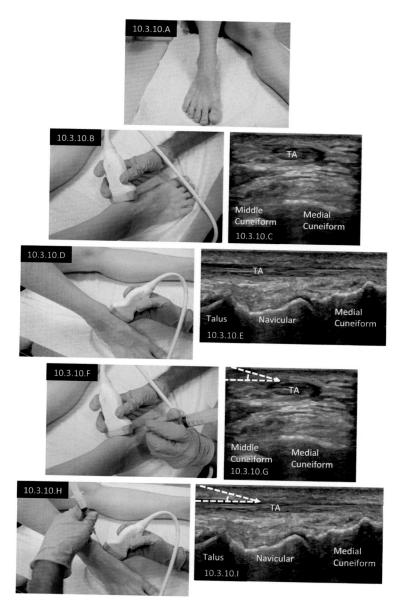

FIGURE 10.3.10 Injections to the TA.

In the supine position the knee can be flexed to 45 degrees and the foot is planted on the couch. The injection can be undertaken with the transducer in a SAX or LAX orientation over the TA, and the needle is introduced using an IP technique between the tendon and the sheath for CSI and into the tendon itself for PRP injections.

Ligament injections

Common ligaments that might require an injection around the ankle include the anterior talo-fibular ligament (ATFL), the calcaneo-fibular ligament (CFL), the deltoid ligament (DL), anterior inferior tibio-fibular ligament (AITFL), plantar fascia (PF) and collateral ligaments (CL) in the digits.

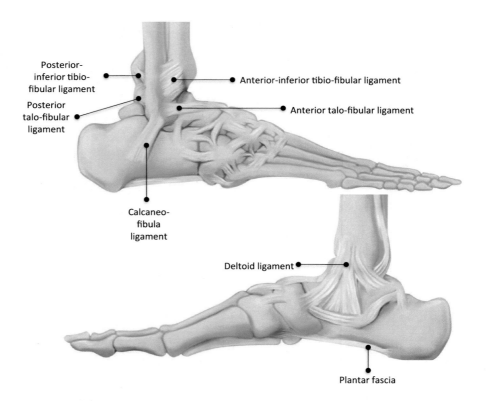

Posterior-inferior tibio-fibular ligament

Posterior talo-fibular ligament

Anterior-inferior tibio-fibular ligament

Anterior talo-fibular ligament

Calcaneo-fibula ligament

Deltoid ligament

Plantar fascia

10.3.11 **Anterior talo-fibular ligament**

Patient position:	For the anterior talo-fibular ligament (ATFL), the patient can be positioned in a supine position with the knee flexed to approximately 45 degrees and if needed this can be further stabilised with a support. The foot should be planted on the examination couch and the ankle is in a slightly internally rotated and plantar flexed position to expose the ligament (Fig. 10.3.11A).
Identifying the anatomy:	The ATFL can be imaged from the dorsal aspect by placing the transducer in a LAX orientation along the fibula and rotating the distal end so that it is parallel to the base of the foot. In this position, the ATFL should appear as a linear structure running from the fibula to the talus (Fig. 10.3.11B and C).
Injections performed:	CSI for pain symptoms. Prolo injections for partial tears.
Recommended transducer:	Linear 6–15 mHz. Hockey stick 8–18 mHz.
Equipment suggested:	*Equipment preparation:* Set 1 for CSI. Set 5 for Prolo. *Syringe:* 3 mL for CSI or Prolo injections. *Medication:* 20 mg triamcinolone (0.5 mL) and 1% lidocaine (1 mL) for CSI. Prolo, 2 mL of a 50:50 mixture of 50% dextrose and 1% lidocaine can be used. *Needle:* 1- to 1.5-inch 25- or 27-gauge needle.
Injection technique:	Viewing the ATFL in the LAX, the needle can be introduced using an IP technique from the distal aspect at an angle of approximately 10–20 degrees (Fig. 10.3.11D and E), with the bevel facing down and the tip aimed for the ligament. For CSI, the needle should be advanced till it is above the ligament, after which a small amount of fluid can be injected to separate the layers before the remainder of the solution can be injected. In contrast for the Prolo injection, the needle tip is passed into the ligament and the solution can be injected using a fenestration technique.

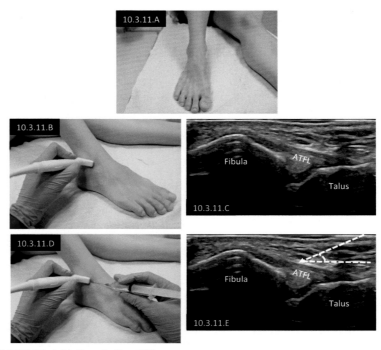

FIGURE 10.3.11 Injections to the ATFL.

In the supine position the knee can be flexed to 45 degrees and the foot is planted on the couch. The foot can be slightly internally rotated to accentuate the lateral structures. The injection can be undertaken with the transducer in a LAX orientation over the ATFL, and the needle is introduced using an IP technique distally. For CSI, the needle is placed above the ligament, and for Prolo it is into the ligament itself.

10.3.12 Calcaneo-fibular ligament

Patient position:	For the calcaneo-fibular ligament (CFL), the patient can be positioned in a supine position with the knee flexed to approximately 45 degrees and stabilised with a support if necessary (Fig. 10.3.12A). The foot can be planted on the examination couch and the ankle is in a slightly internally rotated and inverted position.
Identifying the anatomy:	The CFL can be imaged by placing the transducer along the fibula in a LAX orientation and tracing directly down to the foot (Fig. 10.3.12B and C). It is almost perpendicular to the base of the foot. In this position, the CFL should appear as a slightly convex triangular structure running from the fibula to the calcaneum. It can be accentuated by asking the patient to dorsiflex the foot.
Injections performed:	CSI for pain symptoms. Prolo injections for partial tears.
Recommended transducer:	Linear 6–15 mHz. Hockey stick 8–18 mHz.
Equipment suggested:	*Equipment preparation:* Set 1 for CSI. Set 5 for Prolo. *Syringe:* 3 mL for CSI or Prolo injections. *Medication:* 20 mg triamcinolone (0.5 mL) and 1% lidocaine (1 mL) for CSI. Prolo, 2 mL of a 50:50 mixture of 50% dextrose and 1% lidocaine can be used. *Needle:* 1- to 1.5-inch 25- or 27-gauge needle.
Injection technique:	Viewing the CFL in the LAX orientation, the needle can be introduced using an IP technique from the distal aspect at an angle of approximately 10–20 degrees, with the bevel facing down and the tip aimed for the ligament (Fig. 10.3.12D and E). For CSI, the needle should be advanced till it is above the ligament after which a small amount of fluid can be injected and as the layers separate then the remainder of the solution can be injected. For Prolo injections, once the tip is in the ligament, a fenestration technique can be used to distribute the solution. Care should be taken to avoid injury to the peroneus brevis and longus (PB/PL) tendons that run near the ligament.

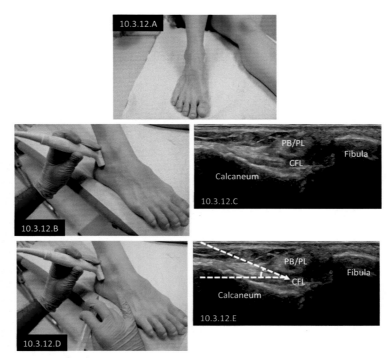

FIGURE 10.3.12 Injections to the CFL.

In the supine position the knee can be flexed to 45 degrees and the foot is planted on the couch. The foot can be slightly internally rotated and inverted to accentuate the lateral structures. The injection can be undertaken with the transducer in a LAX orientation over the CFL, and the needle is introduced using an IP technique distally. For CSI the needle is placed above the ligament, and for Prolo it is into the ligament itself.

10.3.13 **Deltoid ligament**

Patient position:	For the deltoid ligament (DL), the patient can be positioned in a supine side-lying position with foot slightly off the edge of the examination couch (Fig. 10.3.13A). The ankle is placed in a slightly everted to expose the ligament and a towel beneath the lateral malleolus can help stabilise further if needed.
Identifying the anatomy:	The DL can be imaged by placing the transducer in a LAX orientation along the tibia and tracing directly down to the foot. It is almost perpendicular to the base of the foot. In this position, the DL should appear as a triangular structure running from the fibula to the calcaneum (Fig. 10.3.13B and C). The deep and superficial parts may be recognisable.
Injections performed:	CSI for pain symptoms. Prolo injections for partial or degenerative tears.
Recommended transducer:	Linear 6–15 mHz. Hockey stick 8–18 mHz.
Equipment suggested:	*Equipment preparation:* Set 1 for CSI. Set 5 for Prolo. *Syringe:* 3 mL for CSI or Prolo injections. *Medication:* 20 mg triamcinolone (0.5 mL) and 1% lidocaine (1 mL) for CSI. Prolo, 2 mL of a 50:50 mixture of 50% dextrose and 1% lidocaine can be used. *Needle:* 1- to 1.5-inch 25- or 27-gauge needle.
Injection technique:	Viewing the DL in a LAX orientation, the needle can be introduced using an IP technique from the distal aspect at an angle of approximately 20–30 degrees, with the bevel facing down and the tip aimed for the ligament (Fig. 10.3.13D and E). For CSI, the needle should be advanced till it is above the ligament, after which a small amount of fluid can be injected and as the layers separate then the needle can be re-positioned if needed before the remainder of the solution can be injected. For Prolo injections, once the needle tip is in the ligament, it can be injected using a fenestration technique. Care should be taken to avoid injury to the TP tendon that runs near the ligament.

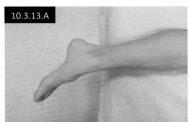

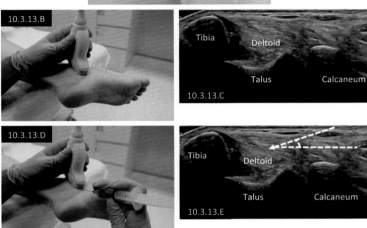

FIGURE 10.3.13 Injections to the DL.

In the supine position the foot hangs slightly off the edge of the couch and the foot can be slightly everted to accentuate the medial structures. The injection can be undertaken with the transducer in a LAX orientation over the DL, and the needle is introduced using an IP technique distally. For CSI, needle is placed above the ligament, and for Prolo it is into the ligament itself.

10.3.14 Anterior-inferior tibio-fibular ligament

Patient position:	For the anterior-inferior tibio-fibular ligament (AITFL), the patient can be positioned in a supine position with the knee flexed to approximately 45 degrees and it can be further stabilised with a support if needed (Fig. 10.3.14A). The foot can be planted on the examination couch and kept in a neutral position.
Identifying the anatomy:	The AITFL can be imaged from the anterior aspect by placing the transducer in a SAX orientation across the distal fibula and tibia. In this position, the AITFL should be seen in the LAX running between the two bones (Fig. 10.3.14B and C).
Injections performed:	CSI for pain. Prolo for minor disruption or partial tears.
Recommended transducer:	Linear 6—15 mHz. Hockey stick 8—18 mHz.
Equipment suggested:	*Equipment preparation:* Set 1 for CSI. Set 5 for Prolo. *Syringe:* 3 mL for CSI or Prolo injections. *Medication:* 20 mg triamcinolone (0.5 mL) and 1% lidocaine (1 mL) for CSI. Prolo, 2 mL of a 50:50 mixture of 50% dextrose and 1% lidocaine can be used. *Needle:* 1- to 1.5-inch 25- or 27-gauge needle.
Injection technique:	Viewing the AITFL in the LAX and the tibia and fibula in the SAX, the needle can be introduced using an IP technique from the medial or lateral aspect at an angle of approximately 10—20 degrees, with the bevel facing down and the tip aimed for the ligament (Fig. 10.3.14D and E). For CSI, the needle should be advanced till it is above the ligament, after which a small amount of fluid can be injected. As the layers separate, the needle can be re-positioned before the remainder of the solution is injected. For Prolo, the needle tip is advanced into the ligament itself and once correctly placed, the solution can be injected using a fenestration technique.

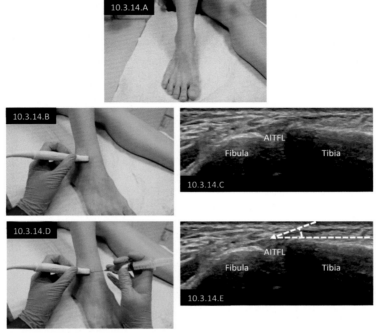

FIGURE 10.3.14 Injections to the AITFL.

In the supine position the knee can be flexed to 45 degrees and the foot is planted on the couch. The injection can be undertaken with the transducer in a LAX orientation over the AITFL, and the needle is introduced using an IP technique. For CSI, the needle is placed above the ligament, and for Prolo it is into the ligament itself.

10.3.15 Plantar fasciitis

Patient position:	For the plantar fasciitis (PF), the patient can be positioned in a prone position with the leg fully extended and the foot hanging off the edge of the treatment couch (Fig. 10.3.15A and D). A rolled towel between the anterior aspect of the ankle and the couch can help stabilise the ankle further.
Identifying the anatomy:	The PF can be traced by placing the transducer in a LAX orientation along the length of the foot and with the calcaneum in view. The PF can be seen attaching on to the calcaneum and typically the medial bundle is thickened. Rotating the transducer 90 degrees, the PF can be seen in the SAX with the calcaneum beneath it (Fig. 10.3.15C and D).
Injections performed:	CSI for pain symptoms. Prolo injections for degenerative changes or intra-substance tears.
Recommended transducer:	Linear 6–15 mHz.
Equipment suggested:	*Equipment preparation:* Set 1 for CSI. Set 5 for Prolo. *Syringe:* 3 mL for CSI or Prolo injections. *Medication:* 20 mg triamcinolone (0.5 mL) and 1% lidocaine (1 mL) for CSI. Prolo, 2 mL of a 50:50 mixture of 50% dextrose and 1% lidocaine can be used. *Needle:* 1.5- to 2.0-inch 25- or 27-gauge needle.
Injection technique:	For CSI, the PF is best viewed in the SAX oreintation and the needle can be introduced using an IP technique from the medial aspect with the bevel facing up. It should ideally be as parallel to the transducer as possilbe. Brought to rest between the PF and fat pad (Fig. 10.3.15E and F), the solution can be injected but care must be taken not to inject into the PF itself due to risk of rupture. Once the tissue planes are separated, the needle can be repositioned if needed, before the injection in completed. For Prolo injections, a similar orientation can be used, but the needle should be directed into the PF (Fig. 10.3.15G) and a fenestration technique is used to distribute the solution. For both interventions, the transducer can be rotated to a LAX oreintation, to observe the needle in an OOP view and identify where in the ligament is situated.

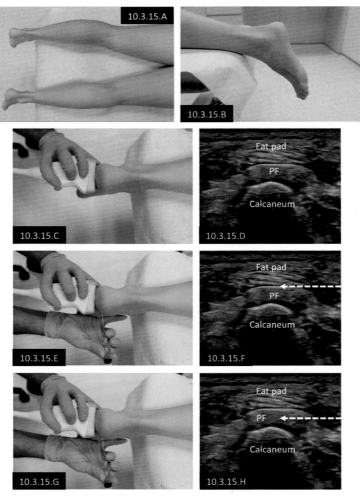

FIGURE 10.3.15 Injections to the PF.

In the prone position the knee is fully extended and foot hangs off the end of the couch. The injection can be undertaken with the transducer in a SAX orientation over the PF, and the needle is introduced using an IP technique from a medial aspect. For CSI, the needle is placed between the ligament and fat pad, and for Prolo it is into the ligament itself.

10.3.16 Collateral ligaments

Patient position:	For the collateral ligaments (CL), the patient can be positioned in a supine position with the knee flexed to approximately 45 degrees and stabilised with a support if needed (Fig. 10.3.16A). The foot can be planted on the examination couch and the ankle is in slightly inverted or everted depending on the position of the ligament to be treated.
Identifying the anatomy:	The CL can be imaged by placing the transducer along the joint in question in a LAX oreintation and identifying the joint in the SAX. The ligament will be seen overlying and traversing the joint (Fig. 10.3.16B and C).
Injections performed:	CSI for pain symptoms. Prolo injections for degenerative changes or intra-substance tears.
Recommended transducer:	Hockey stick 8–18 mHz.
Equipment suggested:	*Equipment preparation:* Set 1 for CSI. Set 5 for Prolo. *Syringe:* 3 mL for CSI or Prolo injections. *Medication:* 20 mg triamcinolone (0.5 mL) and 1% lidocaine (0.5 mL) for CSI. Prolo, 2 mL of a 50:50 mixture of 50% dextrose and 1% lidocaine can be used. *Needle:* 1.5- to 2.0-inch 25- or 27-gauge needle.
Injection technique:	Viewing the CL in the LAX, the needle can be introduced using an IP technique from the proximal aspect at an angle of approximately 10–20 degrees, with the bevel facing down and the tip aimed for the ligament (Fig. 10.3.16D and E). Once within the ligament, the Prolo can be injected using a fenestration technique. An OOP approach can also be used to guide the needle, but visualisation within the ligament may be more challenging (Fig. 10.3.16F and G). For CSI, the needle should be advanced till it is above the ligament after which a small amount of fluid can be injected to separate the layers. The needle can be repositioned if needed at this point, before the remainder of the solution is injected. Care should be taken to avoid injury to any neurovascular structures that lie in close proximity to the ligament.

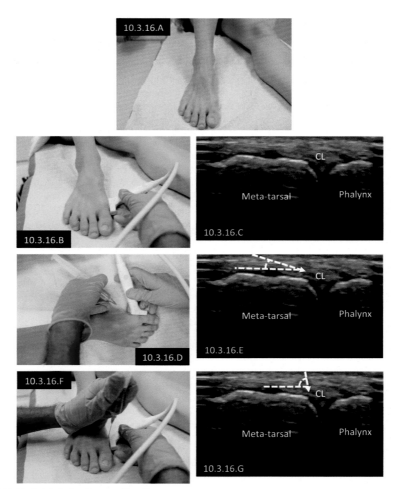

FIGURE 10.3.16 Injections to the CL.

In the supine position the knee can be flexed to 45 degrees and the foot is planted on the couch. The injection can be undertaken with the transducer in a LAX orientation over the CL, and the needle is introduced using an IP or OOP technique. For CSI, the needle is placed above the ligament, and for Prolo it is into the ligament itself.

Nerve injections

Although less commonly undertaken, nerve injections around the ankle and foot include those for the tibial nerve (TN) in the tarsal tunnel and digital nerves in the Morton's neuroma (MorN).

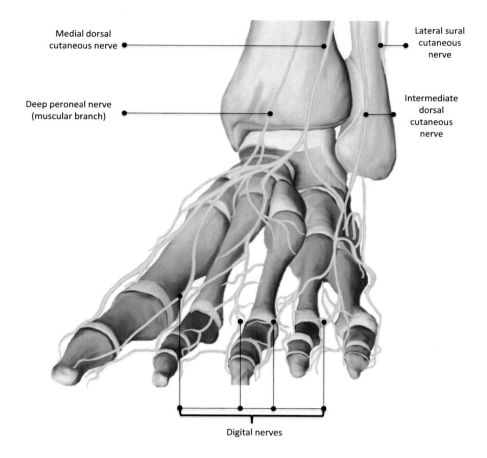

Medial dorsal
cutaneous nerve

Lateral sural
cutaneous
nerve

Deep peroneal nerve
(muscular branch)

Intermediate
dorsal
cutaneous
nerve

Digital nerves

10.3.17 **Tibial nerve**

Patient position:	For the tibial nerve (TN) in the tarsal tunnel (TT), the patient can be positioned in a prone position with the knee extended and the foot hanging off the edge of the examination couch (Fig. 10.3.17.A and B). This exposes the posteromedial aspect of the ankle. If needed a towel can be positioned between the dorsum of the foot and treatment couch to stabilise the ankle further.
Identifying the anatomy:	The TN can be imaged by placing the transducer in a SAX orientation over the posteromedial aspect of the ankle and tracing it distally down to the foot. The Tibia should sit anteriorly and the Flexor Retinaculum (FR) is seen as a ligamentous band running across in a LAX orientation; this represents the upper border of the TT. The TN sits just beneath the FR within the TT (Fig. 10.3.17D).
Injections performed:	CSI for pain and compression symptoms.
Recommended transducer:	Linear 6–15 mHz. Hockey stick 8–18 mHz.
Equipment suggested:	*Equipment preparation:* Set 1 for CSI. *Syringe:* 3 mL for CSI. *Medication:* 20 mg triamcinolone (0.5 mL) and 1% lidocaine (0.5 mL) for CSI. *Needle:* 1.5- to 2.0-inch 25- or 27-gauge needle.
Injection technique:	Viewing the TN in the TT in a SAX orientation, and the FR in the LAX, the needle can be introduced using an IP technique from the posterior aspect at an angle of approximately 10–20 degrees. With the bevel facing down, the tip is aimed for the plane between the FR and neurovascular bundle Fig. 10.3.17E and F. Once the tip is in the space, a small amount of solution can be injected to separate the bundle from the retinaculum and once this is achieved, the needle can be re-positioned if needed before the remainder is injected. Care should be taken to avoid injury to tendons around the medial aspect of the ankle as well as the neurovascular structures themselves. It is important to ensure that the nerve is not touched during the procedure as this will elicit considerable pain.

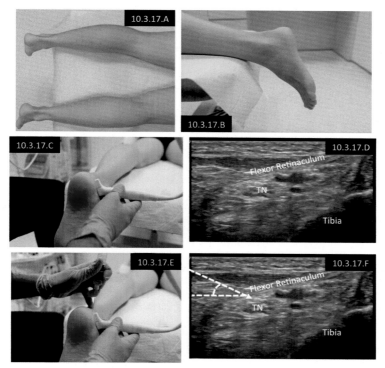

FIGURE 10.3.17 Injections to the TN.

In the prone position the knee is fully extended and foot hangs off the end of the couch. The injection can be undertaken with the transducer in a LAX orientation over the FR, and TN is seen in a SAX view. The needle is introduced using an IP technique from a posterior aspect and is placed between the TN and retinaculum. Care must be taken to avoid injecting into the nerve itself.

10.3.18 Morton's neuroma

Patient position:	For a Morton's neuroma (MorN), the patient can be positioned in a supine position with the knee fully extended and resting on the examination couch. The foot should be in a neutral and relaxed position (Fig. 10.3.18A). A rolled towel between the examination couch and the heel can help stabilise the foot.
Identifying the anatomy:	The MorN can be visualised by placing the transducer in a SAX orientation across the plantar aspect of the foot with the metatarsal heads in view (Fig. 10.3.18B and C). When squeezed, a hypoechogenic swelling can be seen between the metatarsal heads and Mulder's click may be noted. Viewing the webspace in a LAX orientation, a hypoechogenic area can be seen around the level of the metatarsal heads and may be compressible with pressure from the dorsal aspect of the foot (Fig. 10.3.18D and E).
Injections performed:	CSI for pain symptoms.
Recommended transducer:	Linear 6–15 mHz.
Equipment suggested:	*Equipment preparation:* Set 1 for CSI. *Syringe:* 3 mL for CSI. *Medication:* 20 mg triamcinolone (0.5 mL) and 1% lidocaine (0.5 mL) for CSI. *Needle:* 1.5- to 2.0-inch 25- or 27-gauge needle.
Injection technique:	Viewing the MorN with the transducer in a SAX orientation, the needle can be introduced from the dorsum of the foot using an OOP approach with the needle perpendicular to the skin surface (Fig. 10.3.18F and G). The needle tip should be visualised in the space and provided there is no blood on aspiration, then the solution can be injected. Alternatively, an IP approach with the transducer in a LAX position can be utilised. The needle is inserted at 10–20 degrees directly into the webspace distally and directly towards the neuroma (Fig. 10.3.18H and I). Provided there is no blood flow on aspiration, the solution can be injected and will be seen flowing into the hypoechogenic area.

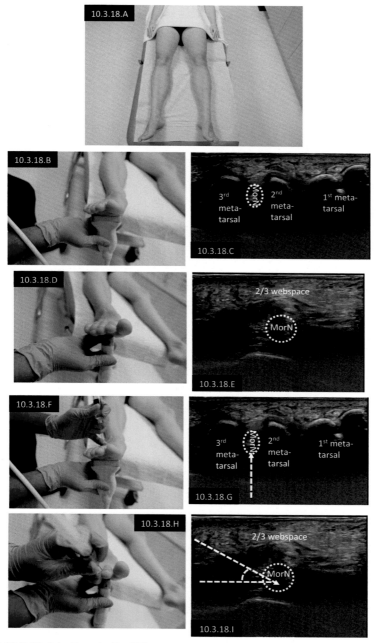

FIGURE 10.3.18 Injections to the MN.

In the supine position the knee is fully extended and foot hangs off the end of the couch. The injection can be undertaken with the transducer in a LAX orientation over the plantar aspect of the foot, and the needle is introduced using an IP technique between the web spaces or an OOP technique from the dorsum of the foot.

Muscle injections

Common injection sites involving muscle around the ankle include the plane between the gastrocnemius and soleus.

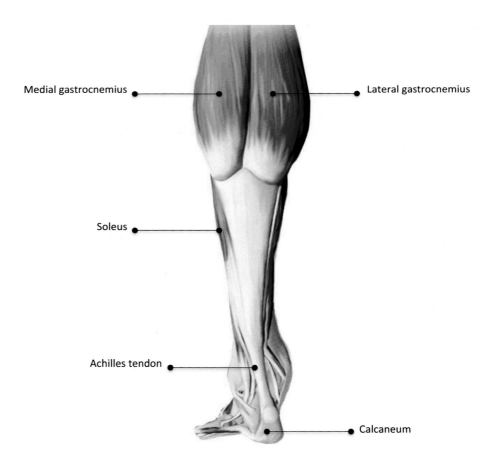

10.3.19 Calf tear

Patient position:	To assess a calf tear (CT), the patient can be positioned in a prone position, with knee fully extended and the foot hanging over the edge of the examination couch (Fig. 10.3.19A). If this position is painful, a rolled towel can offload the calf and also help stabilise the leg.
Identifying the anatomy:	The CT can be visualised by placing the transducer in a SAX orientation to identify the gastrocnemius and soleus muscles. The tear is usually situated in the plane between both muscles and can be seen as a hypoechogenic area (Fig. 10.3.19B and C). Rotating the transducer into a LAX view enables the full extent of the tear to be evaluated (Fig. 10.3.19D and E).
Injections performed:	Aspiration PRP injections for muscle tears.
Recommended transducer:	Linear 6–15 mHz.
Equipment suggested:	*Equipment preparation:* Set 4 for PRP injections. *Syringe:* 10 mL syringe(s) for aspiration. *Needle:* 1.5- to 2-inch 21- or 23-gauge needle. Standard/available PRP preparation.
Injection technique:	Viewing the CT in a SAX orientation, the needle can be introduced using an IP technique from the medial or lateral aspect at an angle of approximately 30–45 degrees depending on body habitus (Fig. 10.3.19F and G). With the bevel facing down, the needle should be brought into the tear and the haematoma can be aspirated. Once this has been completed, the syringe can be swapped to the PRP-containing one and this can be slowly injected into the space. Alternatively, an IP LAX approach at 20–30 degrees can also be used to deliver this treatment, but more intact muscle may need to be traversed to reach the tear and hence there is greater potential for collateral damage (Fig. 10.3.19H and I). Offloading and compression post procedure is important.

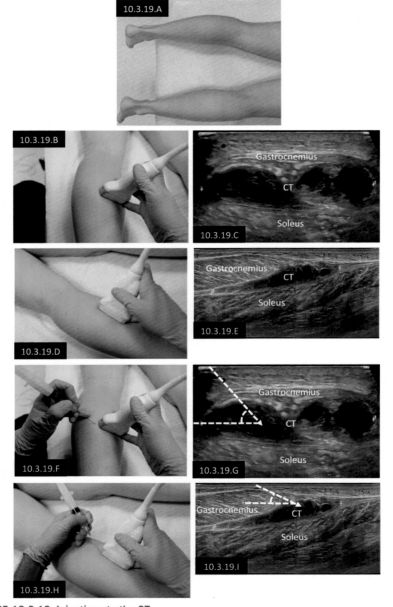

FIGURE 10.3.19 Injections to the CT.

In the prone position the knee is fully extended and foot hangs off the end of the couch. The injection can be undertaken with the transducer in a LAX or SAX orientation over the muscle of the foot, and the needle is introduced using an IP technique in either position. The tear is aspirated, and the PRP can be injected using the same needle.

Notes
(Please use this area to reflect on your procedure and how you can build on these experiences).

Bibliography

1. Hall DE, Prochazka AV, Fink AS. Informed consent for clinical treatment. *Can Med Assoc J.* 2012;184(5):533−540.
2. Kadam RA. Informed consent process: A step further towards making it meaningful!. *Perspect Clin Res.* 2017;8(3):107−112.
3. Shah P, Thornton I, Turrin D, Hipskind JE. Informed consent. In: *StatPearls [Internet].* Treasure Island (FL): StatPearls Publishing; 2020.
4. Shah A, Mak D, Davies AM, James SL, Botchu R. Musculoskeletal corticosteroid administration: current concepts. *Can Assoc Radiol J.* 2019;70(1):29−36.
5. Tay M, Sim H-SS, Eow C-Z, Lor K-HK, Ong J-H, Sirisena D. Ultrasound-guided lumbar spine injection for axial and radicular pain: a single institution early experience. *Asian Spine J.* 2021;15(2):216−223.
6. MacMahon PJ, Eustace SJ, Kavanagh EC. Injectable corticosteroid and local anesthetic preparations: a review for radiologists. *Radiology.* 2009;252:647−661.
7. Kompel AJ, Roemer FW, Murakami AM, Diaz LE, Crema MD, Guermazi A. Intra-articular corticosteroid injections in the hip and knee: perhaps not as safe as we thought? *Radiology.* 2019;293(3):656−663.
8. Klocke R, Levasseur K, Kitas GD, Smith JP, Hirsch G. Cartilage turnover and intra-articular corticosteroid injections in knee osteoarthritis. *Rheumatol Int.* 2018;38(3): 455−459.
9. Maldonado DR, Mu BH, Ornelas J, et al. Hip-spine syndrome: the diagnostic utility of guided intra-articular hip injections. *Orthopedics.* 2020;43(2):e65−e71.
10. Becker DE, Reed KL. Local anesthetics: review of pharmacological considerations. *Anesth Prog.* 2012;59:90−102.
11. Cherobin A, Tavares GT. Safety of local anesthetics. *An Bras Dermatol.* 2020;95(1): 82−90.
12. Kreuz PC, Steinwachs M, Angele P. Single-dose local anesthetics exhibit a type-, dose-, and time-dependent chondrotoxic effect on chondrocytes and cartilage: a systematic review of the current literature. *Knee Surg Sports Traumatol, Arthroscopy.* 2018;26(3): 819−830.
13. De Lucia O, Murgo A, Pregnolato F, et al. Hyaluronic acid injections in the treatment of osteoarthritis secondary to primary inflammatory rheumatic diseases: a systematic review and qualitative synthesis. *Adv Ther.* 2020;37(4):1347−1359.
14. Snetkov P, Zakharova K, Morozkina S, Olekhnovich R, Uspenskaya M. Hyaluronic acid: the influence of molecular weight on structural, physical, physico-chemical, and degradable properties of biopolymer. *Polymers.* 2020;12(8).
15. Zhao J, Huang H, Liang G, Zeng LF, Yang W, Liu J. Effects and safety of the combination of platelet-rich plasma (PRP) and hyaluronic acid (HA) in the treatment of knee osteoarthritis: a systematic review and meta-analysis. *BMC Muscoskel Disord.* 2020; 21(1):224.
16. Mishra A, Collado H, Fredericson M. Platelet-Rich plasma compared with corticosteroid injection for chronic lateral elbow tendinosis. *PM R.* 2009;1(4):366−370.
17. Jain K, Murphy PN, Clough TM. Platelet rich plasma versus corticosteroid injection for plantar fasciitis: a comparative study. *Foot.* 2015;25(4):235−237.
18. Everhart JS, Cole D, Sojka JH, et al. Treatment options for patellar tendinopathy: a systematic review. *Arthroscopy.* 2017;33(4):861−872.

19. Yerlikaya M, Talay Calis H, Tomruk Sutbeyaz S, et al. Comparison of effects of leukocyte-rich and leukocyte-poor platelet-rich plasma on pain and functionality in patients with lateral epicondylitis. *Arch Rheumatol*. 2018;33(1):73−79.

20. O'Connell B, Wragg NM, Wilson SL. The use of PRP injections in the management of knee osteoarthritis. *Cell Tissue Res*. 2019;376:143−152.

21. Eymard F, Ornetti P, Maillet J, et al. Intra-articular injections of platelet-rich plasma in symptomatic knee osteoarthritis: a consensus statement from French-speaking experts. *Knee Surg, Sports Traumatol, Arthroscopy*. 2020. https://doi.org/10.1007/s00167-020-06102-5.

22. Hauser RA, Lackner JB, Steilen-Matias D, Harris DK. A systematic review of dextrose prolotherapy for chronic musculoskeletal pain. *Clin Med Insights Arthritis Musculoskelet Disord*. 2016;9.

23. Dwivedi S, Sobel AD, DaSilva MF, Akelman E. Utility of prolotherapy for upper extremity pathology. *J Hand Surg Am*. 2019;44(3):236−239.

24. Distel LM, Best TM. Prolotherapy: a clinical review of its role in treating chronic musculoskeletal pain. *PM R*. 2011;3(6 Suppl 1):S78−S81.

25. Reeves KD, Sit RW, Rabago DP. Dextrose prolotherapy: a narrative review of basic science, clinical research, and best treatment recommendations. *Phys Med Rehabil Clin N Am*. 2016;27(4):783−823.

26. Sit RW, Chung V, Reeves KD, et al. Hypertonic dextrose injections (prolotherapy) in the treatment of symptomatic knee osteoarthritis: a systematic review and meta-analysis. *Sci Rep*. 2016;6:25247.

27. Sarah M, Chan O, King J, et al. High volume image-guided Injections for patellar tendinopathy: a combined retrospective and prospective case series. *Muscles, Ligaments Tendons J*. 2014;4(2):214−219.

28. Abdelbary MH, Bassiouny A. Ultrasound guided injection in patellar tendinopathy; clinical outcomes of platelet-rich plasma compared to high-volume injection. *Egypt J Radiol Nuclear Med*. 2018;49(4):1159−1162.

29. Klontzas ME, Vassalou EE, Zibis AH, Karantanas AH. The effect of injection volume on long-term outcomes of US-guided subacromial bursa injections. *Eur J Radiol*. 2020;129: 109113.

30. van der Vlist AC, van Oosterom RF, van Veldhoven PLJ, et al. Effectiveness of a high volume injection as treatment for chronic Achilles tendinopathy: randomised controlled trial. *Br Med J*. 2020;370. https://doi.org/10.1136/bmj.m3027.

31. Kim DY, Lee SS, Nomkhondorj O, et al. Comparison between anterior and posterior approaches for ultrasound-guided glenohumeral steroid injection in primary adhesive capsulitis: a randomized controlled trial. *J Clin Rheumatol*. 2017;23(1):51−57.

32. Uppal HS, Evans JP, Smith C. Frozen shoulder: a systematic review of therapeutic options. *World J Orthoped*. 2015;6(2):263−268.

33. Rymaruk S, Peach C. Indications for hydrodilatation for frozen shoulder. *EFORT Open Rev*. 2017;2(11):462−468.

34. Xiao RC, Walley KC, DeAngelis JP, Ramappa AJ. Corticosteroid injections for adhesive capsulitis: a review. *Clin J Sport Med*. 2017;27(3):308−320.

35. Catapano M, Mittal N, Adamich J, Kumbhare D, Sangha H. Hydrodilatation with corticosteroid for the treatment of adhesive capsulitis: a systematic review. *PM R*. 2018;10(6): 623−635.

36. Saltychev M, Laimi K, Virolainen P, Fredericson M. Effectiveness of hydrodilatation in adhesive capsulitis of shoulder: a systematic review and meta-analysis. *Scand J Surg*. 2018;107(4):285−293.

37. Cardone DA, Tallia A. Joint and soft tissue injection. *Am Fam Physician*. 2002;66(2): 283—288.
38. Kruse DW. Intraarticular cortisone injection for osteoarthritis of the hip. Is it effective? Is it safe? *Curr Rev Musculoskel Med*. 2008;1(3—4):227—233.
39. Pekarek B, Osher L, Buck S, Bowen M. Intra-articular corticosteroid injections: a critical literature review with up-to-date findings. *Foot*. 2011;21(2):66—70.
40. Di Matteo B, Filardo G, Presti ML, Kon E, Marcacci M. Chronic anti-platelet therapy: a contraindication for platelet-rich plasma intra-articular injections? *Eur Rev Med Pharmacol Sci*. 2014;18:55—59.
41. Kumar Sahu A, Rath P, Aggarwal B. Ultrasound-guided injections in musculo-skeletal system - an overview. *J Clin Orthop Trauma*. 2019;10(4):669—673.

Index

Printed in the United States
by Baker & Taylor Publisher Services